Bob Samples
Bill Hammond
Bernice McCarthy

4MAT and SCIENCE

toward wholeness in science education

TEACHERS' LABORATORY, INC.
P.O. Box 6480
Brattleboro, VT 05302-6480
Phone (802) 254-3457
FAX (802) 254-5233

©1985 by EXCEL, INC.
200 West Station Street
Barrington, Illinois 60010

Library of Congress 85-071015

ISBN 0-9608992-2-7

Typesetting Graphic Directions, Boulder, Colorado

Printing Corley Printing, St. Louis, Missouri
Coordination Carla Melton

Design and Layout Bob Samples and Bernice McCarthy
Drawings and Artwork Bob Samples
Photographs Bob Samples
Copy Editing Cheryl Charles, Judy Dawson

Dedication

To the spirit of inquiry
To the love of wholeness in knowing
To the honoring of life and living on the Earth planet

To the children of the planet whose experiences
in science will shape the future

To the hope that our children and grandchildren will
find the same inspiration, excitement, and wonder
that we received from those who touched us
with their love for science.

Art Marshal	George Gamow
Eugene Odum	Frank Oppenheimer
Howard P. Odum	Gregory Bateson
John Gustafson	Margaret Mead
E. Lawrence Palmer	Buckminster Fuller
Mary Budd Rowe	David Hawkins
Lewis Thomas	Fritjof Capra
Paul Brandwein	Bill Hazard

Contents

Why Do We Need Learning Styles in Science Education?

The classroom was buzzing. There were two dozen students waiting in anticipation for permission to open the microscope boxes. This was the day they had been waiting for. Seventh grade science was a real departure from the scattered approaches of elementary school. In their first six grades there had been explorations into magnets, insects, stars and digestion, but now there was life science, a whole year on one subject. And the best part of all, there would be microscopes.

A burst of activity followed the teacher's command to open the boxes. Some students fumbled with the cases and locks. Others withdrew the shining instruments at once. Some had the dust shrouds off and were trying to look through the eyepieces. Another group searched for the instruction manual and not finding it, tried to identify the parts from the huge wall chart hanging in front of the room. Three students refused to touch the microscopes at all, left their assigned seats and joined a small group of classmates at a different table trying to determine what all this meant. Two students were trying to light a piece of paper by focusing sunlight through the lens from the eyepiece.

Chaotic? Perhaps. But this scene is repeated at all levels of instruction from preschool through college whenever students are excited about learning. They revert to unique, individual behaviors because their fascination for the material causes them to forget momentarily the ubiquitous notion that we must all approach learning in the same way. Sometimes the instructional materials are rocks, worksheets, art supplies, magnets or jigsaw puzzles. Whatever the material, any sudden rush into this diverse array of behaviors on the part of students has been traditionally judged as poor teacher organization. After all, instruction is supposed to be a tidy and orderly process.

To try to add tidiness and order, many teachers have resorted to high amounts of programmed instruction, unrelenting rules, and multiple guidelines. The belief persists that there must be rules, procedures, and steps to follow to insure control. Often these teachers make the claim that science is done in a single controlled way. We do not raise the question of whether rules and order are useful and even helpful, but we do raise serious questions as to whether rules and order should serve to support science instruction or to dominate it.

Rather than honor the diversity that flourishes in our classrooms, teachers often try to homogenize the behavior of the students into a narrow range of controlled interactions. Much of this tendency is born of the assumption that learning takes place in a well understood and orderly fashion. Research shows that this is not so. Learning is a remarkably personal matter.

A closer look at the classroom described above provides the insight that the students were using individual approaches effective for them. Each student has learning preferences. The seeming chaos was, in reality, an expression of the diversity in how people learn. **These learning styles, although different, are equally effective in the acquisition of skills, knowledge and critical thinking.** In the "open" setting created by the teacher allowing free access to the microscopes, the possibilities were high that the students would express their own learning style preferences.

Teachers can establish rules to insure proper use of equipment and yet honor the diversity of the students' learning preferences as they actually explore science.

In the pages that follow we will explore how the recognition and honoring of specific learning styles in this seemingly chaotic diversity can be guided in a systematic way to insure growth, and we will examine the ways exposure to other styles of learning offers the gifts of flexibility and fluency to students and teachers alike.

Learning is a remarkably personal matter.

The most beautiful and profound emotion that we can experience is a sense of the mystical. . .it is the dower of all true art and science.

Albert Einstein

The Why of Science: A Search For A Definition

Two decades ago the definition of science as a human activity was generally expressed as a methodology. The **Scientific Method,** as it was popularly called, had a sequence of steps that were often listed as follows:

1) Recognize the existence of a problem.
2) Define the problem.
3) Suggest multiple working hypotheses.
4) Test the hypotheses.
5) Formulate a solution.
6) Test the solution again.
7) Formally state the solution.

The importance of this series of steps in American science education was profound. Its origins were most recently from American philosopher, John Dewey. In an essay about the analysis of a complete act of thought, Dewey formulated these steps as a kind of circuit diagram of internally consistent logic. So powerful was the influence of this method that for two decades materials in science education were framed around these steps. Many who sponsored the seven step method did not bring Dewey's insight to the enterprise.

We may have to give up the pretense of objectivity and admit that what we probably really need is an enlightened subjectivity.

Margaret Mead

Dewey proposed that school be a miniature workshop and a miniature community. The teaching method should be trial and error, and seen as continuous growth. "In a sense, the schools can give us only the instrumentalities of mental growth; the rest depends upon our absorption and interpretation of experience. Real education comes after we leave school."[1]

Those educators who took up Dewey's steps discarded his imperative that experience be integrated with all learning and thus the practice became lock-step, wooden and ritualistic. The sequence of science experiences for children often followed the progression of this recipe. It is no wonder that students were less than enchanted with the enterprise. For many science became a lifeless, routinized form of filling in the blanks. In such settings, students were never introduced to the adventures of the unknown. They never had the opportunity to confront some self-discovered mystery of the world in which they lived.

The rote progression of experiences replaced the rote memorization of facts. Some of the more common sequences of instruction fell into patterns like these:
TELL-TEXT-TEST
TELL-DEMONSTRATE-TEXT-TEST
TELL-TEXT-DEMONSTRATE-TEST
These were the primary delivery cycles of instruction in science. In some more "enlightened" school systems students might experience the sequence:
TELL-TEXT-DEMONSTRATE-LAB-TEST. And "lab" was many times a canned experience that was in reality a demonstration the students performed rather than the teachers.

Science is thinking very hard about things
in as many ways as you can.

Gregory Bateson

None of these progressions truly represented science but that discovery was still on the horizon. The federal government began to listen to the complaints of an army of practicing scientists who were adamant about bringing education in their fields closer to the reality of the actual practices of science. They wanted the excitement and sense of adventure to be emphasized as well as the thoroughness and discipline. Science began to be defined in ways that wedded knowledge, methodology and the spirit of the human quest. For twenty years scientists, teachers and psychologists explored ways to make science teaching more effective. The result was the emergence of what affectionately has been called the "alphabet" science programs:

AAAS, ESCP, ESS, SCIS, SAPA, BSCS, PSSC, IPS, etc.[2]

The commitment was to honor the processes of inquiry as well as the fundamental content of science.

Today because of remarkable new information about the human brain-mind system, we can take the results of the "alphabet" effort and judge from a wider spectrum about the contributions of science to our lives and to our ways of thinking.

Science is a far wider enterprise than following a closed system recipe.

In this book we are committed to explore ways of teaching science that contain contemporary views of what science is and how the workings of the brain-mind system honor those views. Science has escaped the narrowness once attached to it by those who thought of it as a body of facts and laws. Science is a far wider enterprise than the following of a closed system recipe. Science is a uniquely human endeavor with its explorations approaching the issues of the quality of life and the very possibilities of survival on this planet. It is an activity that calls forth all the qualities that make us uniquely human.

Emotion, values, excitement, commitment and concern are joining the arena of discipline, responsibility and integrity. Science is a human activity. It is a whole brain process which engages the widest possible range of the mind's capacities. Science is **sensing**, questioning, **dreaming**, analyzing, **synthesizing**, investigating, **guessing**, tearing apart, **putting together**, measuring, **feeling**, and looking into the unknown, all with a sense of excitement.

Science is a quest to know and it involves the whole of our mind, body, and soul.

Science is a whole brain process that engages the widest possible range of the mind's capacities.

Some of my most complex problems in physics could
be solved by children playing in the streets.
They have not given up the ways of knowing
I lost long ago.

J. Robert Oppenheimer

With all we now know, the following definition of science seems appropriate:

Science is a way of knowing about ourselves and the world of which we are a part. It is a sense of wonder and a childlike way of knowing paradoxically characterized by integrity, discipline, and responsibility. The goal of the enterprise is to provide awareness, understanding, and wisdom for a greater harmony between ourselves and our world.

Few have addressed this issue better than Dr. Mary Budd Rowe[3], a noted science educator. Dr. Rowe contends that science education is composed of four parts. These are

1) Ways of knowing
2) Actions/applications
3) Consequences
4) Values

She then points out that three of the four are generally left out of science instruction. We teach primarily for number one, ways of knowing, leaving out actions/applications, consequences, and values. Ultimately these issues address the great "so what" question, the **meaning** question, so vital to our young people today. Dr. Rowe claims that if we leave out these components, we will be disenfranchising a great many of our students.

This is a claim that the authors endorse heartily.

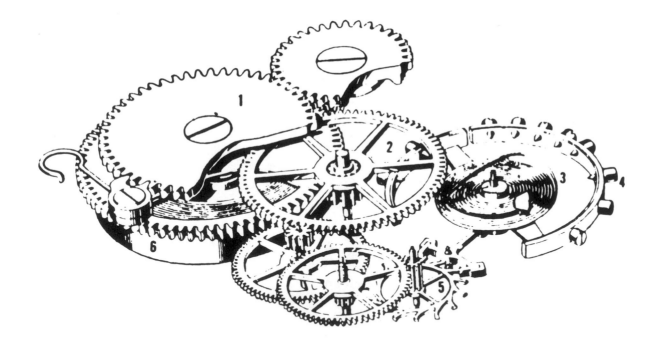

What Makes 4MAT And Science A Natural?

In this century the history of science education has been marked by rather remarkable shifts in perspective. We started with the notion that science was a body of facts and information, a repository of the products of inquiry. Gradually that view was replaced by a commitment to the processes and methodologies of investigative inquiry. In the emerging new science curricula of the 1950s and 1960s, there was a shift toward process approaches. Instruction shifted from lecture-demonstration to an emphasis on inquiry skills.

This shift established the tone of science education throughout the 1970s. But somewhere late in the decade difficult times seemed to fall on science education. An amalgam of factors overlapped and sifted into schools throughout the nation. The following characteristics seemed present:

1) Time spent in science instruction began to drop. (Many attributed this to the "back to the basics" factor.)
2) Funding resources for education at all levels, community, state and federal, became diminished.
3) Students who could choose began to choose options other than science.

The third of these factors was of major concern to many of those committed to science education. What had happened to science that seemed to be driving students to other electives? After all, the approach had shifted from the lecture-demonstration methods to hands-on involvement which should have been more appealing to many. Yet interest and enrollment in science continued to wane. What had happened to the capacities for natural interest and fascination inherent in science? Somehow the hands-on, high involvement approaches were not enough. Teachers sensed that although much had been gained, something was still missing. Looking back now, it seems that most approaches were still biased toward the formal, rigorous side of science. The act of discovery was confined to the concepts being studied and wholeness was still missing. The approach over-emphasized verification rather than the sense of mystery and invention, so central to the drama of true science.

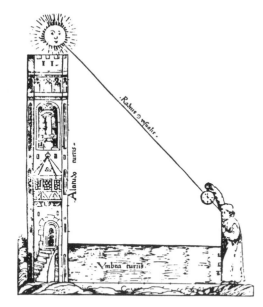

What Brain-Mind Research Tells Us

The recognition of the commitment to teaching the whole child is central to **The 4MAT System** approach and the emerging research in the brain-mind system. We will be returning to the **4MAT** approach, but first we need to explore the findings from brain-mind research, learning modalities, and the growing importance of learning styles in education.

Neural research has told us much about learning modalities; learning styles has told us how to apply it.

Research about the brain-mind system is continuing to experience a golden age. For nearly two hundred years there was a quiet time in neural research related to how the brain works. The quiescence of this period was largely due to the lack of technological ways to study intact, undamaged brains. Most neural information came to us as a result of research performed on damaged brains; accidents, internal lesions, and strokes.

Less than twenty years ago, a group of neurosurgeons established a surgical procedure that brought remarkable new information to the field. Dr. Roger Sperry[4] and Dr. Joseph Bogen[5] developed a radical new surgical technique to reduce the effects of massive epileptic seizures. The technique consisted in the surgical separation of the two sides of the human cortex. This procedure was the result of the conclusion that the seizures originated due to uncontrollable feedback between the hemispheres of the brain, resulting in an epileptic lapsing into a helpless loss of faculties. Since these were life threatening events, the radical surgery was attempted.

The results were remarkable. The patients were clearly relieved of the major symptoms of epilepsy. They were returned to a normal existence. But were they normal? As the researchers conducted the battery of post recovery tests, they were amazed to discover the patients used each side of their now separated brains in distinctly different ways. The research was quite complex and detailed, but it rapidly became clear that the right side of the brain did very different kinds of processing than did the left.

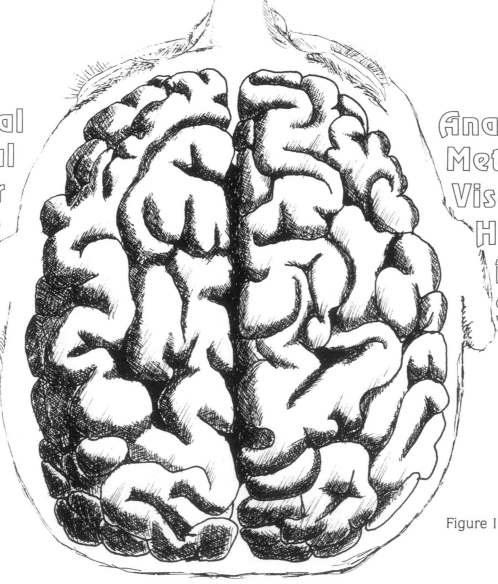

Logical
Rational
Linear
Reductive
Time-Ordered
Sequential

Analogic
Metaphoric
Visual-Spatial
Holistic
Proliferative
Simultaneous

Figure I

In general the left brain processes experience in logical, linear, analytic, and reductive ways. It seems to quest for detail and specificity. The right half of the brain tends to process experience in more holistic, metaphoric, and analogic ways. It is biased toward larger systems of relationships, and is more involved in visual and spatial reasoning.

In a whole brain, one not separated by surgery, these functions would be interwoven in such a way that their distinction might go undetected. The impact of the "split-brain" research was that it provided the first definitive data concerning the different ways the brain processes experience. Before the work of Sperry and Bogen, psychologists, educators and neuroscientists could only address this reality intuitively. With these data, scientists began to look at clearly separate ways of processing experience in learning, memory, and thinking.

Since this breakthrough, there have been innumerable advances in the technologies of monitoring functions in the brain-mind system. Biofeedback, CAT-scans, PET-scans, magnetic resonance devices, and thermal monitoring have all entered the arena as tools of exploration of this awesome frontier. For those of us in education, insights and guidance into the fabric of teaching and learning are expanding each day. The following sections will explore some of the impact of brain-mind research and related fields. In particular, we will return to the interrelatedness of the brain-mind system and instruction.

This will be discussed in the section **4MAT** And The Whole Brain.

Learning Modalities

The brain-mind system has certain parts that are activated by different kinds of input from the outside world. Visual input through forms and images provides excitation to different regions of the brain than do written words. It is clear that both images and written words are visual, but the inputs are for such different purposes that the brain has developed the ability to sort them out. The same is true with sound input. Music and unstructured sound activate very different areas of the brain than spoken language. In other words, the brain is notably selective in regard to where excitation takes place in the face of varying inputs. Some researchers have treated these as physiological "habits" that stretch back across eons of adaptation. Although this selectivity exists in the brain-mind system, there is a remarkable ability to double-up and piggyback the processing task if there is damage to the "first choice" areas. These two facts seem contradictory, but in the mysterious workings of the brain-mind system they are not. The brain seems to be able to specialize and generalize in both physiological and psychological ways. That is, if the brain is physically damaged in a place where a particular form of processing is localized (speech, for example), then other areas of the brain begin to pick up the processing task from the damaged areas. Similarly if the brain has developed preferential "habits" of processing which are not physiologically but psychologically based, then we can intentionally excite different parts of the brain through instruction.

What this means is we can intentionally employ teaching techniques to excite the brain-mind system in a holistic way. We call the framework for this intentional variation of brain-mind excitation **Learning Modalities.**[6]

Although there are a myriad of ways to excite the brain-mind system, we have compressed the list to the following:

> Symbolic/abstract modalities
> Visual modalities
> Auditory modalities
> Kinesthetic modalities
> Synergic modalities

Each of these is based on research in the neurosciences, research that supports brain excitation pattern differences in response to physiological input. In addition they are chosen to represent the widest array of excitation patterns which are expressed when output is elicited.

Symbolic/Abstract Modalities are those activated when input and output involve the use of abstract codes such as letters, numbers, or symbols. The three R's represent this mode. Research justifies separation of these modalities into two subsets; one relates to linguistic symbols, and the other to symbols of mathematics.

Visual Modalities are those which are activated by input or output using visual/spatial expression such as art, sculpture, mapping, graphics, and other forms of visually dominated expression.

Auditory Modalities are those which are activated by input or output in the use of patterned sound. Speech is the earliest formal expression of these modalities but music, song and rhythmic awareness are also primary.

Kinesthetic Modalities are those which are excited by input and output involving movement. Movement is fundamental to basic human understanding, and patterned movement and dance are the core of much human learning.

These four modalities are clearly an abbreviated list, but they represent a basic set familiar to educators. And more importantly, their usefulness is workable. It is possible to systematically give students varying experiences in these four modalities, and research has shown that attention to these four can result in excitation of large areas of the brain.

We acknowledge the importance of a fifth modality. This modality is called **Synergic** and it involves the combinations of those listed above. Use of the **Synergic Modality** results in an empathic relatedness between self, other people, and natural systems.

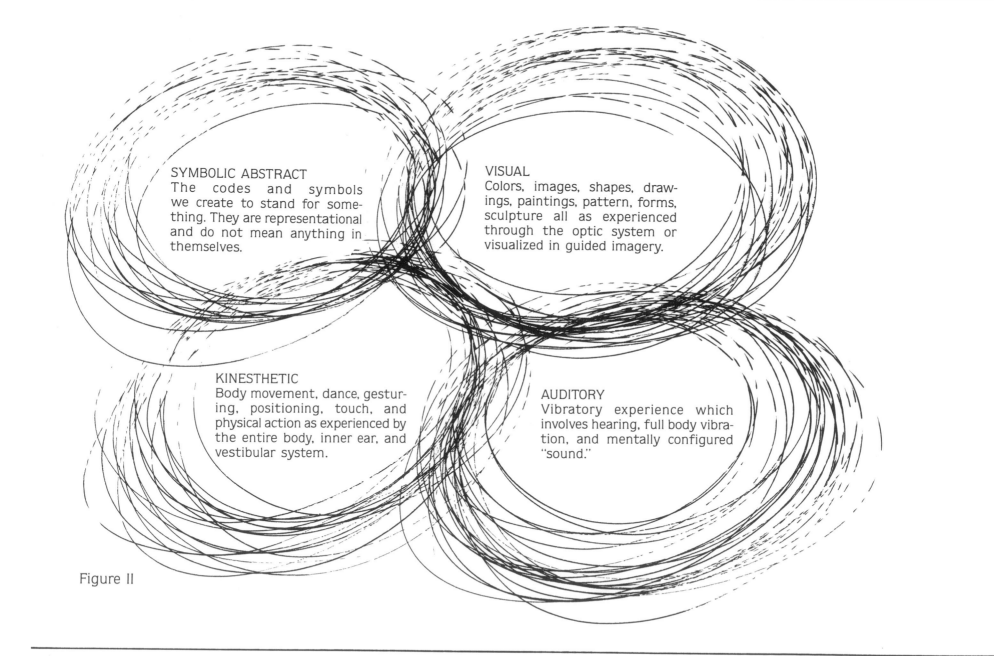

SYMBOLIC ABSTRACT
The codes and symbols we create to stand for something. They are representational and do not mean anything in themselves.

VISUAL
Colors, images, shapes, drawings, paintings, pattern, forms, sculpture all as experienced through the optic system or visualized in guided imagery.

KINESTHETIC
Body movement, dance, gesturing, positioning, touch, and physical action as experienced by the entire body, inner ear, and vestibular system.

AUDITORY
Vibratory experience which involves hearing, full body vibration, and mentally configured "sound."

Figure II

Learning Styles

Just as we have determined that the human brain processes experience differently in its different parts, so too does the brain-mind system choose different ways of processing experience as a whole. This means the entire brain-mind system has within it as much variability as each of its parts seems to exhibit to researchers. The sum total of the brain mind's ways of perceiving experience coupled with the individual's favorite way of acting on that experience encompasses what had come to be known as learning style.

Over the years, many researchers have explored differences in styles of learning. Some explored factors such as preferences for feeling, sensing, thinking, and intuiting. Others choose personality characteristics, and still others aspects of personal choice in reflecting and acting. Affective and cognitive styles were also central to emerging models. As is often true, the research grew along parallel paths.

A strong sense of cohesion was added to the learning style research in 1971 when David Kolb[7] of the Weatherhead School of Management at Case Western Reserve University in Cleveland proposed an elegant learning styles rationale. Kolb proposed that the major factors responsible for learning styles were two dialectics of opposites. One ranged from **concrete to abstract,** and the other from **active to reflective.** The poles of each of these pairs represent the ends of a continuum that stretches between them. Kolb's work provided the basis for Bernice McCarthy to synthesize the results of various learning styles researchers into one cohesive model.

Kolb had arranged two axes, one horizontal, the other vertical, as shown in Figure III.

By superimposing the composite learning style descriptions on the Kolb model, McCarthy was able to accommodate the major elements of nearly all the other learning styles researchers into a single synthesis. The result is **The 4MAT System.**[8]

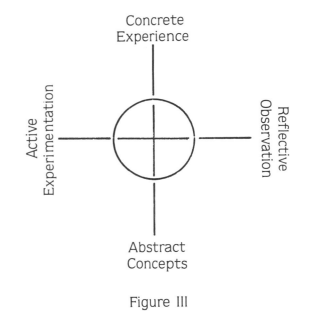

Figure III

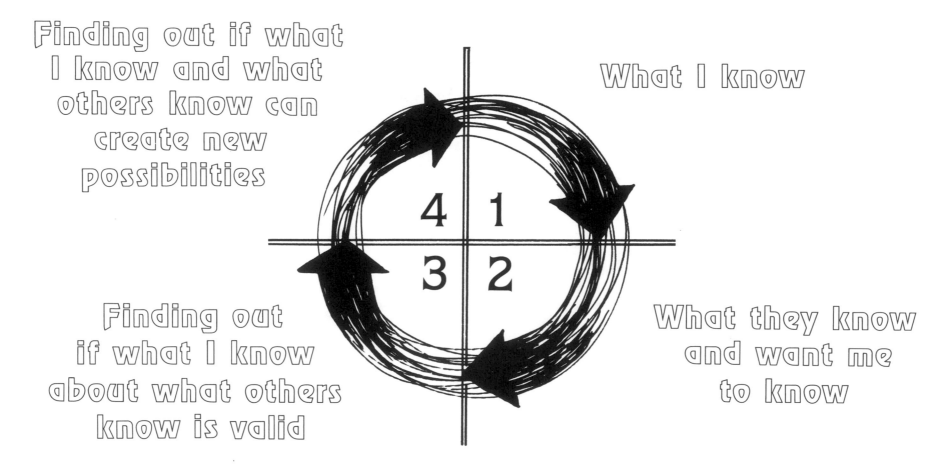

Finding out if what I know and what others know can create new possibilities

What I know

Finding out if what I know about what others know is valid

What they know and want me to know

4 1
3 2

Figure IV

4MAT: A Comprehensive Approach

The 4MAT System combines the characteristics of the two sets of dyads shown above in Fig. III in such a way that they represent combinations of preferences. This combination of preferences results in a pair of tendencies that describe four quadrants. In **The 4MAT System,** each of these quadrants then becomes a different Learning Style. Each quadrant and its pair of descriptors describes a set of tendencies and preferences that different people would exhibit in their attempts to learn and to teach.

Several paths may be explored in describing the **4MAT** approach. One is to describe the system as a personal statement about how an individual sees learning emerging in his or her life. Another way is to describe **4MAT** in a clinical third person way as though it were happening to another person—a hypothetical learner. Still another way would be to invite the reader to take action on a series of tasks and see what happens in each of the four quadrants. And of course we could describe **4MAT** in a way that would provide guidance on how to apply the approach to others in other settings.

Happily it is the way of **4MAT** to include all these parts into a single, comprehensive description. Thus you do not have to choose.

Let us turn to **4MAT** through the personological approach. This exploration is stated in the first person.

Quadrant One—What I know
Quadrant Two—What they know and want me to know
Quadrant Three—Finding out if what I know about what others know is valid
Quadrant Four—Finding out if what I know and what others know can create new possibilities

Quadrant One Personological

In the all important arena of values there is also an evolving perspective. At first my exploration is bonded to the values I possess. It is a case of my relating myself to the environment to forge values, beliefs, and attitudes that work for me in a personal way. I require time to clarify my views and compare them with the views of others.

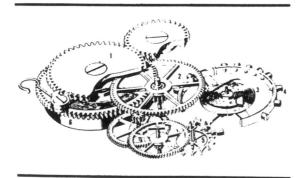

Quadrant Two Personological

Once I make my affective world clear to myself, I am ready to move into the arena of public values, beliefs, and attitudes. Here I can compare the external frameworks of society with the personal frameworks I have established for myself. At this point, information about how others respond to the external environment is useful and extending to me.

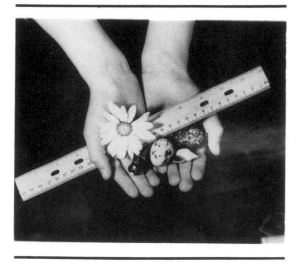

Quadrant Three Personological

Now to test the options. Armed with my own personal values, beliefs and attitudes and an array of cultural options, I can now find out how the two match. If my personal attributes have been honored through the process of learning, then an authentic adoption and adaptation of the sets of values, beliefs, and attitudes of others will occur in my life.

Quadrant Four Personological

With a perspective born of a synthesis of
personal attributes and cultural preferences, I am now
ready to try my new "self" on the world at large.
The being that I was at the beginning has evolved so
as to represent a larger realm of options than I had as
I began the learning cycle. Nowhere have I
experienced a denial of where I was at the outset.
Rather I have been validated in the presence of new
and affirming experiences.

. . .let me offer the following rough map for innovation in science—you start by loving a subject.

E.O. Wilson

When we look at the **4MAT** cycle from this personal perspective, it provides a clear expression of the "meaning in education" concerns of Mary Budd Rowe. (See page 13.)

We are destined to insure fragmentary knowing, superficial at best, unless we truly understand that learning must encompass the completeness and wholeness represented by a cycle of exploration. Superficiality and fragmentation have injured students and teachers in all subject areas, but especially in science.

By now you have discovered one of the basic biases of **4MAT.** Although each learner has a "home-base", a preference for learning, we hope that by using the **4MAT** learning cycles approach, students and teachers will gain more confidence and strength in the kinds of experiences found in quadrants they typically may not prefer. In fact we hope that all students and teachers will develop an awareness of the advantages inherent in each way of knowing.

The real goal is to have learners develop a rounded kind of competence, allowing them to recognize excellence in all forms of human wisdom.

4MAT Descriptors Synthesized by McCarthy[9]

Quadrant One Learner

Integrates experience with the "self."
Seeks meaning, clarity, and integrity.
Needs to be personally involved. Seeks commitment.
Exercises authority with participation and trust.
Learns by listening and sharing ideas. Values insight thinking.
Works for harmony. Absorbs reality.
Perceives information concretely and processes it reflectively.
Interested in people and culture. Divergent thinkers who believe in their own experience, and excel in viewing concrete situations from many perspectives. Model themselves on those they respect.

Strength: Innovation and imagination. They are idea people.
They function through social integration and value clarification.
Goals: Self-involvement in important issues, bringing unity to diversity.
Favorite question: WHY?
Careers: Counseling, personnel teaching, organizational development, humanities, and social sciences.

Quadrant Two Learner

Forms theories and concepts. Seeks facts and continuity.
Needs to know what the experts think.
Seeks goal attainment and personal effectiveness.
Exercises authority with assertive persuasion.
As leaders, they are brave and protective. Learns by thinking through ideas. Values sequential thinking, needs details.
Forms reality. More interested in ideas than people.
Perceives information abstractly and processes reflectively.
Critiques information and collects data.
Thorough and industrious. Reexamines facts if situations are perplexing. Enjoys traditional classrooms.
Schools are designed for these learners.
Functions by thinking things through and adapting to experts.

Strength: Creating concepts and models.
Goals: Self-satisfaction and intellectual recognition.
Favorite question: WHAT?
Careers: Natural sciences, math, research, planning departments.

Quadrant Three Learner

Practices and personalizes.
Seeks usability, utility, solvency, results.
Needs to know how things work. Exercises authority by reward and punishment. Leads by inspiring quality, the best product.
Learns by testing theories in ways that seem most sensible.
Values strategic thinking, is skills oriented. Edits reality.
Perceives information abstractly and processes it actively.
Uses factual data to build designed concepts, enjoys solving problems, resents being given answers. Needs hands-on activities.
Restricts judgement to concrete things, has limited tolerance for fuzzy ideas. Need to know how things they are asked to do will help them in real life. Function a great deal through inferences drawn from their bodies, their kinesthetic selves. They are decision makers.

Strength: Practical application of ideas.
Goals: To bring their view of the present into line with future security.
Favorite question: HOW DOES THIS WORK?
Careers: Applied sciences, engineering.

Quadrant Four Learner

Integrates experience and application.
Seeks hidden possibilities, excitement.
Needs to know what can be done with things.
Exercises authority through common vision.
Leads by energizing people. Learns by trial and error, self-discovery. Seeks influence and solidarity. Enriches reality.
Perceives information concretely and processes it actively.
Is adaptable to change and even relishes it. Likes variety and excels in situations calling for flexibility.
Tends to take risks, at ease with people, sometimes seen as pushy.
Often reaches accurate conclusions in the absence of logical justification. Functions by acting and testing experience.

Strength: Action, carrying out plans.
Goals: To make things happen, to bring action to concepts.
Favorite question: IF?
Careers: Marketing, sales, action-oriented managerial jobs, education, social professions.

Teaching to all four Learning Styles

Remember, each of the four learning style types has a quadrant, or place where he or she is most comfortable, a place where success comes easily.

The Innovative Learners, those who fall in Quadrant One, prefer to learn through a combination of sensing, feeling, and watching. They require personal meaning.

The Analytic Learners, those who fall in Quadrant Two, prefer to learn through a combination of watching and thinking about concepts. They require facts and information.

The Common Sense Learners, those who fall in Quadrant Three, prefer to learn by thinking through concepts and trying things out for themselves, by doing. They require action and must test out facts and information.

The Dynamic Learners, those who fall in Quadrant Four, prefer to learn by doing, sensing, and feeling. They require applying and extending what they learn.

The 4MAT System is designed so all four types of learners are comfortable some of the time. As they gain more experience with other preferences, they tend to establish comfort with styles once avoided.

IF I AM A QUADRANT ONE LEARNER, I LIKE TO
have personal experience with what I learn,
experience personal meaning in what I learn,
learn about things I value and care about,
express my beliefs, feelings and opinions, and
understand how what I learn affects me.

IF I AM A QUADRANT TWO LEARNER, I LIKE TO
get new and accurate information,
deal in facts and right answers,
know what the experts think,
formulate theories, models and plans, and
have things exact and accurate.

IF I AM A QUADRANT THREE LEARNER, I LIKE TO
  do things,
  have ideas clear and understandable,
  find out how things work,
  test theories in the real world, and
  make things useful.

IF I AM A QUADRANT FOUR LEARNER, I LIKE TO
  connect things together,
  do things that matter in life,
  teach other people what I know,
  turn people on and take some risks, and
  make what is already working, work better.

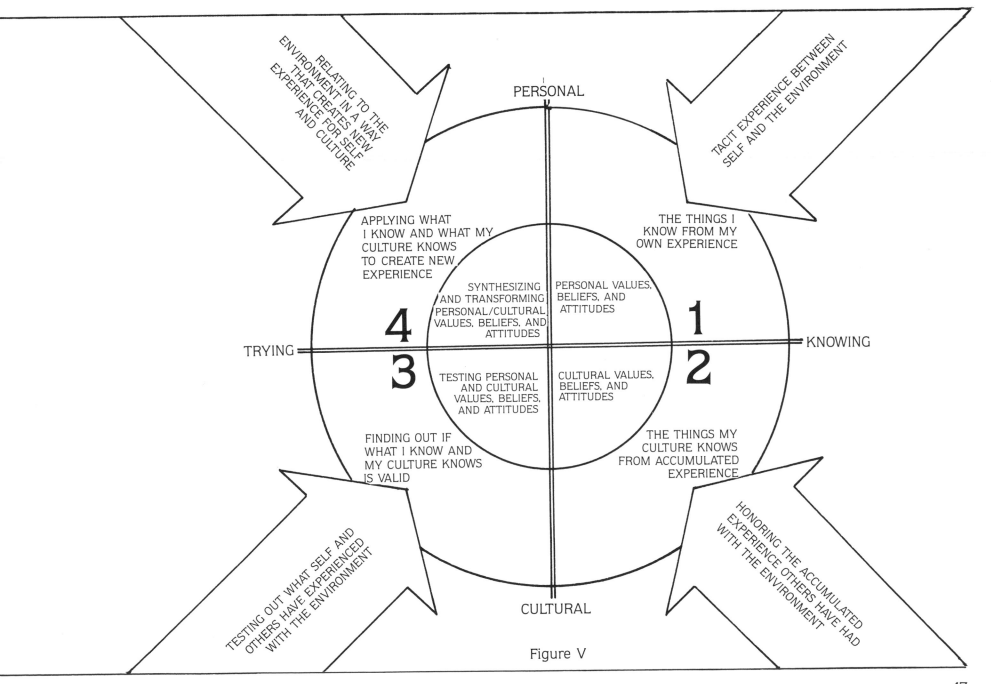

RELATING TO THE ENVIRONMENT IN A WAY THAT CREATES NEW EXPERIENCE FOR SELF AND CULTURE

TACIT EXPERIENCE BETWEEN SELF AND THE ENVIRONMENT

PERSONAL

APPLYING WHAT I KNOW AND WHAT MY CULTURE KNOWS TO CREATE NEW EXPERIENCE

THE THINGS I KNOW FROM MY OWN EXPERIENCE

SYNTHESIZING AND TRANSFORMING PERSONAL/CULTURAL VALUES, BELIEFS, AND ATTITUDES

PERSONAL VALUES, BELIEFS, AND ATTITUDES

4

1

TRYING

KNOWING

3

2

TESTING PERSONAL AND CULTURAL VALUES, BELIEFS, AND ATTITUDES

CULTURAL VALUES, BELIEFS, AND ATTITUDES

FINDING OUT IF WHAT I KNOW AND MY CULTURE KNOWS IS VALID

THE THINGS MY CULTURE KNOWS FROM ACCUMULATED EXPERIENCE

TESTING OUT WHAT SELF AND OTHERS HAVE EXPERIENCED WITH THE ENVIRONMENT

HONORING THE ACCUMULATED EXPERIENCE OTHERS HAVE HAD WITH THE ENVIRONMENT

CULTURAL

Figure V

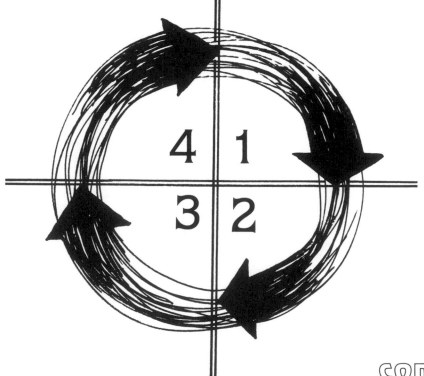

Learning must encompass the completeness of a full cycle of exploration

If you start somewhere other than Quadrant one, then the whole cycle must be completed and then continued again to Quadrant four

4MAT And The Whole Brain

4MAT is a blending of research in the neurosciences and the combined fields of psychology, personality, and motivation to form an innovative approach to instruction. This blending is accomplished through attention to both Learning Modalities and Learning Styles.

Instructional approaches are offered in **4MAT** which are basically designed to honor both right and left brain processing in each of the learning style quadrants. This increases the likelihood of addressing the whole brain as well as the personal learning style preferences. **4MAT** is fundamentally a holistic approach based on neurophysiology, psychology, and pedagogy.

Recently Dr. Siegfried Streufert[10] at the Pennsylvania State University College of Medicine concluded that being able to approach problems in a multidimensional way and avoid being rigidly locked into a single system of thought is more important than I.Q. in insuring success. His work, based on the concept of cognitive complexity, suggests strongly that instruction can address the issues of multi-modal and multi-style thinking in a direct way.

4MAT addresses this perspective exactly. Not only does **4MAT** allow each student to experience success in his or her specific learning style, **it insures that each student will expand his or her personal repertoires of thinking.**

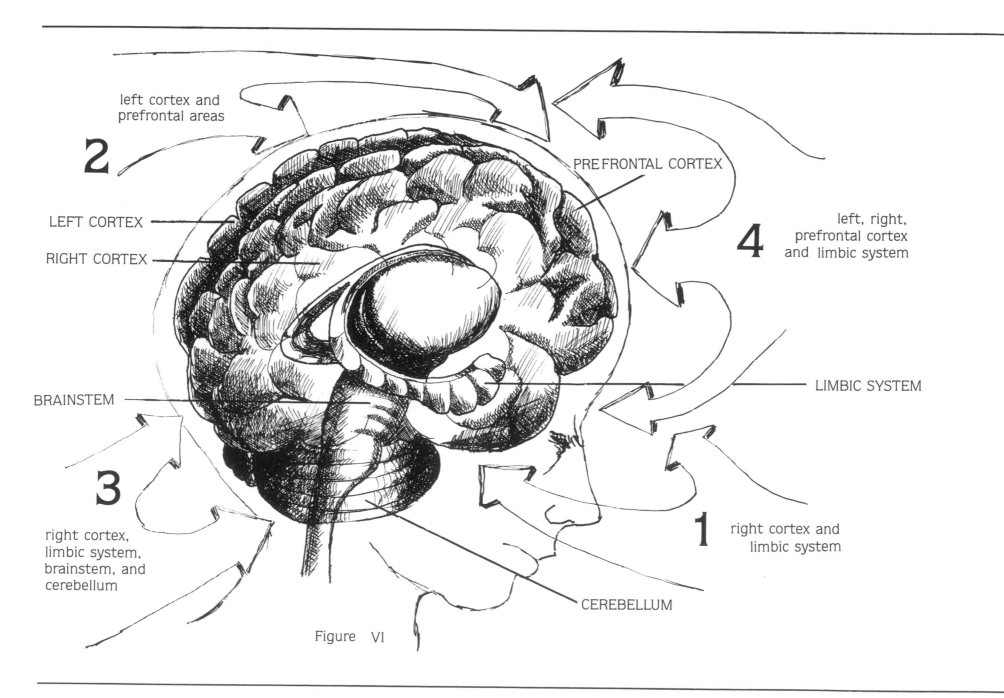

left cortex and
prefrontal areas

2

LEFT CORTEX

RIGHT CORTEX

PREFRONTAL CORTEX

4

left, right,
prefrontal cortex
and limbic system

LIMBIC SYSTEM

BRAINSTEM

3

right cortex,
limbic system,
brainstem, and
cerebellum

1

right cortex and
limbic system

CEREBELLUM

Figure   VI

Research indicates that parts of the brain, once thought to be devoted to functions other than learning, are actually central to learning. The limbic (midbrain) system active in emotional functions and the brainstem system, active in survival functions, are supported by the activities of Quadrants One and Four. In other words, experiences that attempt to bring about a balance between the internalization of felt experience and the development of the courage to act as exemplified by **4MAT** Quadrants One and Four actually engage the whole brain-mind system in a more consciously active way. Generally contemporary education spends little time in honoring these areas of learning and development, and science education in particular tends to ignore them.

Recent research related to the limbic system confirms the connection between Quadrants One and Four and limbic functions. In particular, there is a growing body of information[11] to support the claim that virtually **all** decisions are made in this area of the brain. A person cannot proceed toward positive learning if not at ease and free from fear. Any rational processes related to cortical function are thus monitored by the limbic system. If there is not a sense of comfort and balance between attitudes, beliefs, emotions and values, then the cortex is not permitted free access to the decision making process.

Future science education will be as successful as our ability to invite whole minds into learning. We cannot ignore the realities of tomorrow. We cannot afford an educational experience for the children of today that forces their minds into limited ways of knowing.

A free society is born of free minds.

THE 4MAT SYSTEM: A CYCLE OF LEARNING

Human beings **perceive** experience and information in different ways.
Human beings **process** experience and information in different ways.
The combinations formed by our own perceiving and processing techniques form our unique learning styles.

There are four major identifiable learning styles.
They are all equally valuable.
Students need to be comfortable about their own unique learning styles.

Quadrant One Learners are primarily interested in personal meaning. Teachers need to honor their demand for a Reason in learning.
Quadrant Two Learners are primarily interested in the facts as they lead to conceptual understanding. Teachers need to **Give Them Facts** that deepen understanding.
Quadrant Three Learners are primarily interested in how things work. Teachers need to **Let Them Try It.**
Quadrant Four Learners are primarily interested in self discovery. Teachers need to **Let Them Teach It To Themselves and Others, and Apply It in Life.**

All students need to be taught in all four ways, in order to be comfortable and successful part of the time while being stretched to develop other learning abilities.

All students will "shine" at different places in the learning cycle, so they will learn from each other.

The 4MAT System moves through the learning cycle **in sequence**, teaching in all four modes and incorporating the four combinations of characteristics. The sequence is a natural learning progression.

Each of the four learning styles needs to be taught with both right and left brain processing techniques. This is where Learning Modalities come in.

The left mode dominant students will be comfortable half of the time, and will learn to adapt the other half of the time.

The right mode dominant students will be comfortable half of the time, and will learn to adapt the other half of the time.

The development and integration of all four Learning Styles and the development and integration of both right and left brain processing skills through Learning Modalities should be a major goal of education.

Students will come to accept their strengths and learn to capitalize on them, while developing a healthy respect for the uniqueness of others, and furthering their ability to learn in alternative modes without the pressure of "being wrong."

The more comfortable we are about who we are the more freely we learn from others.

We cannot afford to force the minds of today's children into educational experiences that limit their ways of knowing.

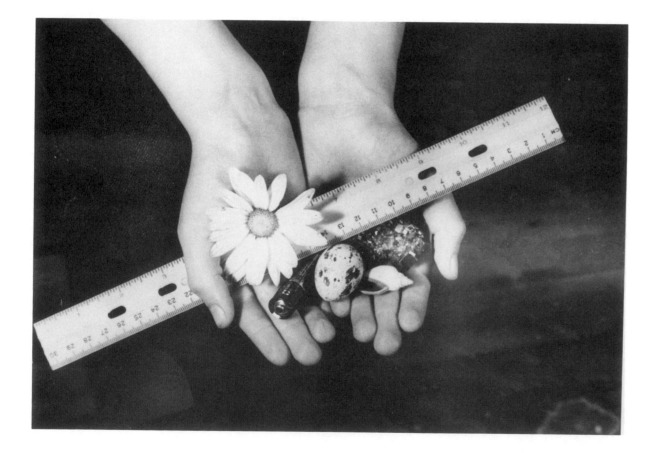

How To Teach Science From The 4MAT Model

Both the teacher and the student co-share experiences when Learning Modalities and Learning Styles are used. Sometimes it means teachers are guiding learning in ways that seem foreign to them. When following **The 4MAT System** some teachers find some of the assignments we recommend unusual and even pointless. An example might be the recommendation to use guided imagery when teaching the laws of motion in physics. For some physics teachers it might seem questionable to ask the students to sit back, relax and close their eyes—and then have the same teacher ask the students in slow, modulated tones to try to see things moving in space and hitting each other. It may also seem to be a waste of time to ask students to try to formulate how the laws of motion have affected their lives.

However our experience, and that of hundreds of teachers, confirms that the imagery and personal meaning assignments enhance learning and are a vital necessity to nearly half of the students we teach.

Our experience working with teachers at all levels confirms there are often discrepancies between a teacher's personal preference for learning and the way that teacher teaches. This seems particularly true at secondary and college levels. Most teachers teach the way they were taught. It is an enlightening revelation to many teachers to discover the multitude of options they have in regard to learning. These same teachers, having experienced the possibilities of personal growth inherent in a cycle of learning that capitalizes on all four Learning Styles, are delighted in turn to extend the same options to their students. When teachers grow, students grow.

If we are to grow in harmony with our world,
we must educate young people who are
scientifically literate,
who believe in their experience
while honoring the wisdom of the experts,
who dare to know and yet to imagine,
and who dialogue with two voices,
one born of imagination and metaphor,
the other born of careful analysis and exactness.

A great deal of talent is lost in this world
for the want of a little courage.

Sidney Smith

Students also have interesting reactions to **4MAT**, because the **4MAT** assignment sequences contain invitations to perform in ways seldom used in standard instruction. The teaching of science has become stereotyped to mean careful, exact, fact filled instruction. So the change is difficult for many students, especially the older ones. It is vital to exhibit skills of encouragement and good natured prodding to get the process started.

A survey in which elementary school students were asked to close their eyes and visualize a scientist at work produced some revealing results. Virtually all the students saw a male figure in white. The visualized scientist was working with chemical apparatus or a microscope. Only rarely did they see a person out of doors. The stereotypes the students possess may well rest so firmly in the limbic filters mentioned earlier that many students, unless assisted, may have great difficulty gaining access to a richer, fuller vision of science.

The vision of a scientist questing for pattern provides personal meaning to the process of discovery (Quadrant One). Without a larger vision the student may never discover the aesthetics of order created and extended by the Quadrant Two scientist. Lost may be the methodical delight of the Quadrant Three scientist, who tests abstractions for their grounding in reality. And finally a narrow perspective of science may never open the door from the laboratory to allow structured knowledge to reach into life, the commitment of the Quadrant Four scientist.

When science is experienced outside stereotypes, there is a great range of feeling and emotion in the process. For if science is in fact an exploration of the unknown, there is a quality of exhilaration as well as uncertainty that accompanies the process.

It is important for both teacher and students to show curiosity in the face of the tentativeness that emerges in times of unfamiliarity and uncertainty. As minor as it seems, this form of courage can grow to a robust kind of self confidence in surprisingly little time.

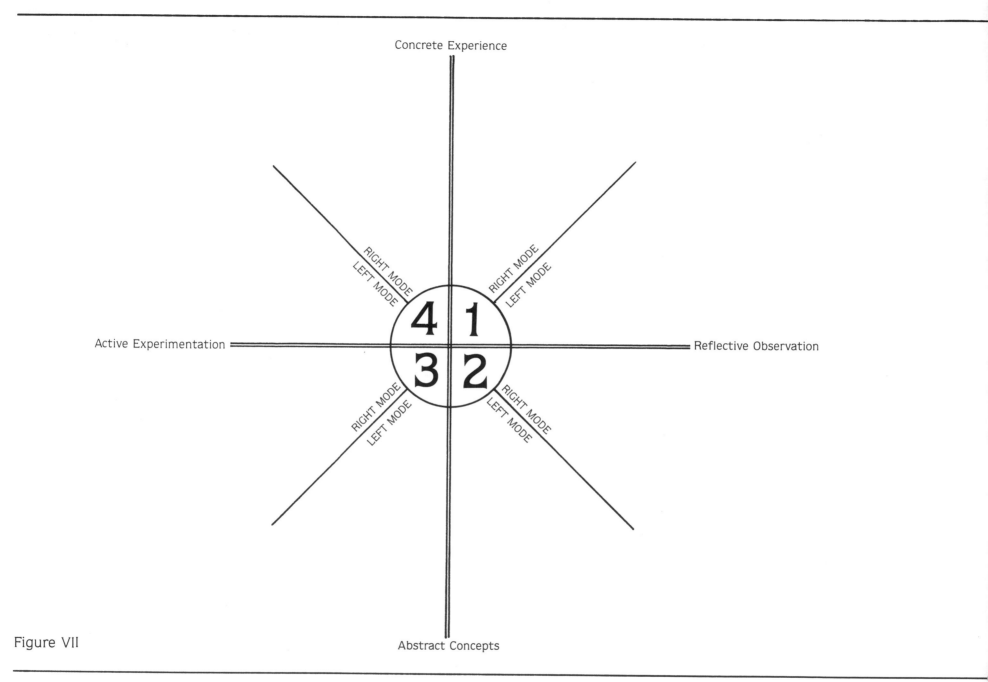

Figure VII

How To Teach About Rocks With 4MAT

We will now take a topic and run it through **The 4MAT System** to see how it works in practice.

The topic is **Rocks.**
The level is **Upper Elementary or Middle School.**

All the students in the class will receive the same assignments. But all will not find each assignment best suited to her or his learning style.

4MAT is designed so each of the four learning styles gets to shine 25% of the time.

4MAT honors both Right and Left brain modalities.

So the **cycle of assignments** built around the topic "Rocks" will include all learners' strengths.

ROCKS: Quadrant One

Personal meaning is central to the learning experiences of Quadrant One Learners. They also require group discussion where the teacher and classmates listen to what they have to say. They want to be recognized in a way that honors their personal history and experience.

We can wed the characteristics of the Quadrant One Learner to a whole brain approach. It is here that we also pay attention to the **Learning Modalities:** Symbolic/Abstract, Visual, Auditory and Kinesthetic.

Each of the modalities can be either left or right brained. However, a good rule of thumb is to consider the Symbolic/Abstract modality, (codes of letters, numbers and symbols) far more likely to be left mode forms, and to consider the Visual, Auditory and Kinesthetic Modalities far more likely to be right mode forms.

If we choose to use art, music or movement, we honor the Visual, Auditory and Kinesthetic realms which favor right mode processing.

If we choose language use in speech, writing or mathematical notation, it is likely to be a left mode experience.

But please note that the fully facile whole brain can create poetry with the symbols of language and art with the codes of numbers. Yet for us these left and right mode distinctions are useful in creating an instructional mood for a more complete kind of education.

ROCKS: Quadrant One

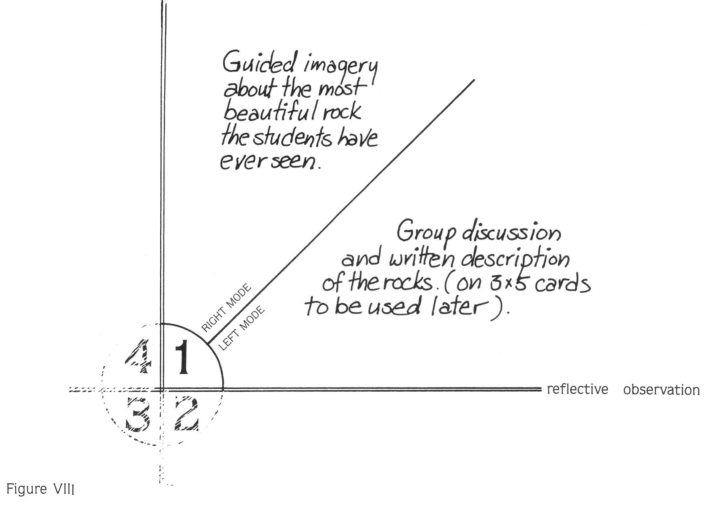

concrete experience

Guided imagery about the most beautiful rock the students have ever seen.

Group discussion and written description of the rocks. (on 3×5 cards to be used later).

RIGHT MODE

LEFT MODE

reflective observation

Figure VIII

ROCKS: Quadrant One

Right Mode:
Close your eyes and try to recreate the most beautiful rock you have ever seen in the natural world...Try to remember everything you can about the setting...Note the colors and smells...Imagine what the rock feels like...Pretend the rock can talk and tell you about its history. Listen carefully to its story so you can remember it. You will have two full minutes from now.

Left Mode:
Gather together in groups of four and tell each other all you can about your rocks and the stories of your rocks. After the discussion write down a description of your rock on a 3 x 5 card. Include its color, shape, if it is heavy or light, and its texture (is it bumpy, smooth, jagged...). Save these descriptions for later.

Quadrant One is the place where students are validated for being the person they are, and for bringing the world of their experience into our presence.

COMMENTARY ON QUADRANT ONE

The use of guided imagery and the intentional solicitation of other senses clearly produces right brain excitement. The invitation to discuss and organize thought and then to write a description of the rock assures left brain processing. The drawing out of the students' past experience from the child's past and their experiences fulfills a basic preference of Quadrant One Learners. The imagery itself provides a concrete experience for the child to use if memory tends to have faded. The two minutes at the end of the imagery offers even more time for reflection, a real need of both Quadrant One and Two Learners. (And we believe, a necessity for all learners.) Additional time for reflection comes in the group discussion when the students listen to the stories of others.

For most teachers, and for science teachers in particular, Quadrant One is a most delicate one. There is the strong tendency to insert adult observation and teacher talk. Here we want to create a base of legitimacy for the students' experiences, so the urge to inject our expertise or to over question must be resisted. . .There will be plenty of time later as we move through the **4MAT** cycle, to ask questions and direct instruction. In Quadrant One, the teacher is a witness and a silent guide. The teacher encourages, legitimizes and honors discussion, but does not guide it. There are no "right" answers in Quadrant One activities. It is the place where students are validated for being the person they are and for bringing their world of experience to our presence.

Over guidance or manipulation of the discussion in Quadrant One can turn it into an encounter group or a kind of sensitivity session. Teachers who are themselves Quadrant One Learners are most likely to overreact in this way. Because sometimes their attentiveness to the qualities of concrete experience and reflective observation in themselves adds a kind of expertise that can invite an overworking of the experiences.

ROCKS: Quadrant Two

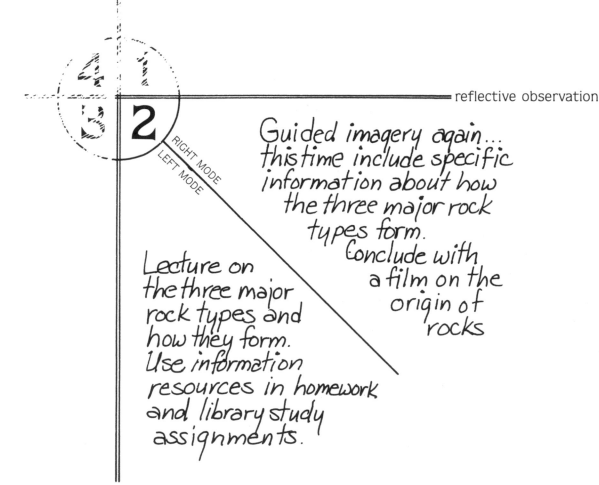

reflective observation

RIGHT MODE
LEFT MODE

Guided imagery again... this time include specific information about how the three major rock types form.
Conclude with a film on the origin of rocks

Lecture on the three major rock types and how they form. Use information resources in homework and library study assignments.

abstract concepts

Figure IX

ROCKS: Quadrant Two

Quadrant Two Learners have a strong preference for facts, order, and sequential thinking. They value ideas and are markedly interested in people who are "experts" in their field. Their quest is for answers and they need objectivity. For them, an internal consistency in abstract ideas and concepts provides an aesthetic form seldom experienced by other learners. This section of the unit, while written for all students, will especially appeal to them.

Right Mode:
Again the teacher might use a guided imagery, but this time moving into more objective facts, taking the students into far more depth than the personal experiences of Quadrant One.

"We are going to use guided imagery again to continue our exploration of rocks. This time we will explore some of the ways rocks are formed. Close your eyes and relax . . . Some rocks form in the waters of oceans or lakes . . . Most often they are made of particles of mud or sand that drift to the bottom of the water . . . Try to visualize this process. After long periods of time, they may become crushed and fastened together by different materials . . . Think back and see if this is how you think your favorite rock formed."

Continue with the imagery:
"Other rocks formed deep within the earth and were once melted or molten . . . Sometimes these rocks come to the surface in volcanos . . . Try to imagine rocks pouring out of the earth . . . Some never make it to the surface and end up cooling down deep in the earth. These rocks often have crystals and large grains of different materials . . . Try to imagine them forming."

"Some rocks are near the deep heat sources of the volcanos. These rocks are baked and even remelted by the heat . . . Try to visualize rocks being melted and twisted by other molten rocks flowing by."

"Keep your eyes closed, review the three kinds of rock origins: those formed in water . . . those formed by molten rock inside the earth and at its surface . . . and those formed by being crushed, twisted, and melted by other rocks . . ."

Now ask the students to open their eyes and watch a film you have ready on the origin of rocks.

ROCKS: Quadrant Two

Quadrant Two is the place where students explore
the grace and aesthetics of structure,
order and sequential thinking.

Left Mode:
Now lecture on the three major rock types and their origins. Provide samples of sedimentary, igneous and metamorphic rocks they can handle. Give the formal definitions of rock types and have the students verify the definitions in the encyclopedia and the dictionary. Be sure to ask the students to find the root meanings of the three words.

COMMENTARY ON QUADRANT TWO

As mentioned, the use of guided imagery is a direct stimulator of the right brain. Using a film at the end of the imagery continues the right brain involvement, but it begins to bring the left mode into play through the introduction of specific information. The right mode is also involved as the students see and handle samples of each rock type. This multi-sensory involvement in the exploration of the rock types brings an aspect of "whole-brainedness" to the experience. One could add to the multi-sensory by asking the students to smell the rocks and perhaps tap them with a hammer to hear how differently they sound. Finally the left mode dominates when the definitions are provided and the students check the standard reference sources.

Quadrant Two approaches represent classical science instruction. However, for some teachers only the left mode activities seem worthwhile, while the imagery and tactile activities would seem less important, and in some instances, ridiculous. This is the tragedy of science education in many cases. We have created an image of scientists as one dimensional detached people. Yet the working scientists we know are all vitally aware of senses. They recognize subtleties in odor among soils, plants, and rocks. They are sensitive to nuances of color, shade, and texture. And they have a rich sense as to heft and weight. We heartily support science education that develops the skills we find in working scientists. It is this multi-dimensional science experience that **4MAT** nurtures.

Quadrant Three is the place where the students experimentally test the concepts they have learned in the proving ground of the real world.

ROCKS: Quadrant Three

The learners who are most at home in Quadrant Three also enjoy the abstract ideas of Quadrant Two. But rather than being fascinated by the theoretical as the Quadrant Two Learners are, what they enjoy is the opportunity to test ideas out. Whereas the Twos are into reflective observation, the Threes want action, and will flourish when activity is required. They are the engineering corps and the maintenance force of any classroom.

Aside from providing the learning style most appropriate for these learners, Quadrant Three approaches are important in the evaluation process. They include writing or completing worksheets, questions at the ends of chapters, graphing, mapping and so on. But Quadrant Three activities also include tinkering, testing, and personally using the material learned. Here is what a continuation of the rock unit into Quadrant Three might look like.

ROCKS: Quadrant Three

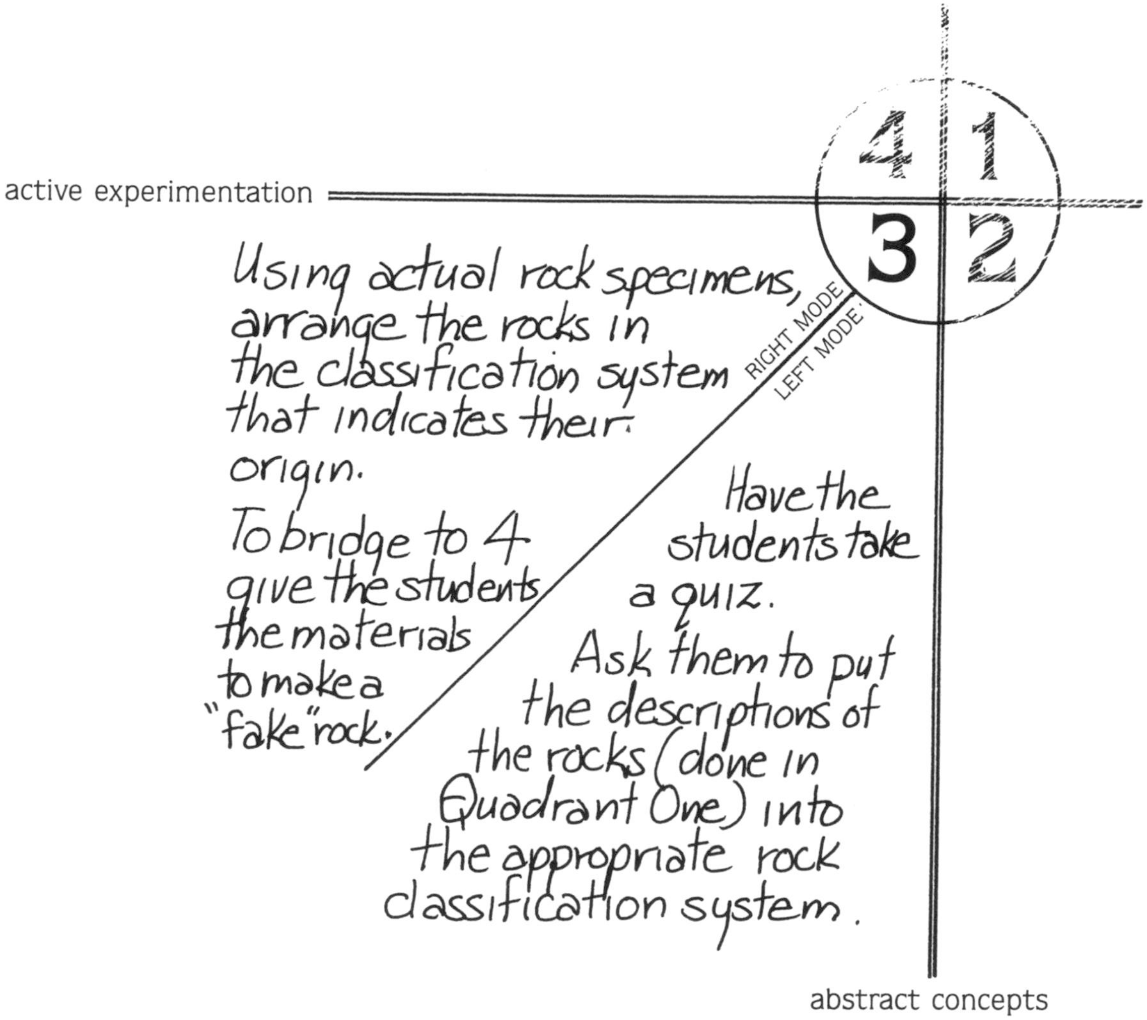

active experimentation

4 | 1
3 | 2

RIGHT MODE
LEFT MODE

Using actual rock specimens, arrange the rocks in the classification system that indicates their origin.

To bridge to 4 give the students the materials to make a "fake" rock.

Have the students take a quiz. Ask them to put the descriptions of the rocks (done in Quadrant One) into the appropriate rock classification system.

abstract concepts

Figure X

ROCKS: Quadrant Three

Left Mode:
Answer the questions on the quiz provided and do the worksheets on rock formations and how they are classified. A second assignment might ask that students use the descriptions they wrote on the 3 × 5 cards during the Quadrant One (Left Mode) activity. In groups of three or four, have the students classify their original rock descriptions as near as they are able.

Right Mode:
Using actual rock specimens arrange the rocks into classification systems based on their origins. Match these with the 3 × 5 cards and discuss the things remembered from the guided imagery. Encourage them to close their eyes and see if they can reconstruct the way they saw the rocks forming. Offer each student a half a cup of sand and ask them to try to make a synthetic rock over the weekend. Emphasize their "fake rock" must represent the same processes that are involved in the formation of rocks in nature.

COMMENTARY ON QUADRANT THREE

The linking of the high action components with the concepts of Quadrant Two forms the main thrust for these Type Three activities, but an important new dimension has been added. The students have been asked to use the concept in ways that encourage imagination, resourcefulness and experimentation. They have been asked to "tinker", the favorite pastime of Quadrant Three Learners. And so the process pattern formed by The **4MAT** Cycle as we have moved through Quadrants One, Two and Three alternates from right to left mode, from the known to the unknown, from the subjective to the objective.

Quadrant One: The Self, pulled in by past experiences, experiences highly subjective and personal.

Quadrant Two: The Known, the objective, the body of knowledge, the linking of things named with their origins.

And now in Quadrant Three: First, the left mode activities to make sure the basics are mastered, then back to the Self, the personal use, the right mode that blends the gestalt with the specific, engages the divergent, and encourages the creation of usefulness unique to each student. In addition, the manipulation of symbols involved in writing and reading in the quiz and worksheets aid the left mode experience; while the right mode is involved in the visual, kinesthetic, and imaging components.

Quadrant Four learners seek more than the classroom as they quest to make a difference in the world and try to wed what they have learned with the fabric of their lives.

ROCKS: Quadrant Four

The Quadrant Four learners want to know how all this works in the real world, and their passion extends quickly beyond the limits of the classroom. They tend to see applications far beyond the immediate lessons. Their quest for personal involvement in what they learn is crucial to them because of their preference for Concrete Experience. Their need for action, based on their Active Experimentation preference, is their chief motivation. The Quadrant Four Learner needs to make a difference. So activities written in Quadrant Four need to extend learning beyond the classroom. But not only for the Quadrant Four Learners, but for all learning style types.

These learners must extend what they learn into their own lives. If they cannot, they simply have not learned it.

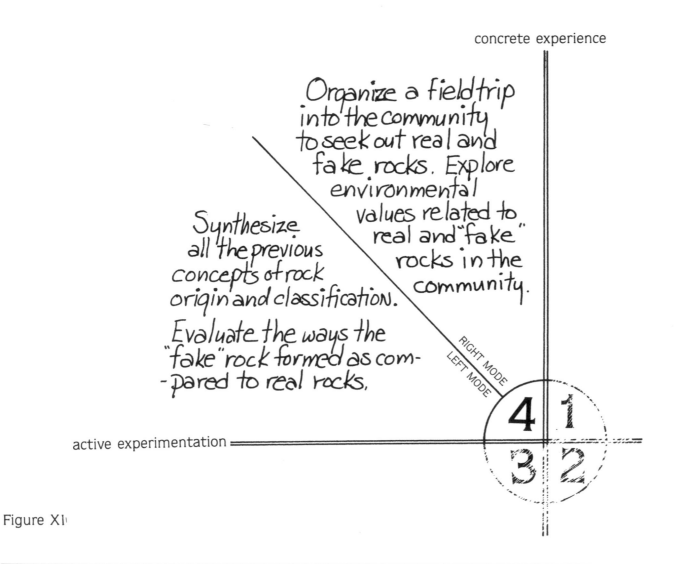

concrete experience

Organize a field trip into the community to seek out real and fake rocks. Explore environmental values related to real and "fake" rocks in the community.

Synthesize all the previous concepts of rock origin and classification.

Evaluate the ways the "fake" rock formed as com- -pared to real rocks.

RIGHT MODE
LEFT MODE

active experimentation

4 1
3 2

Figure XI

Left Mode:
The final assignment given in Quadrant Three is an excellent one to begin a synthesis activity that reviews all the major concepts of the unit. We have prepared the students to review all the key concepts of the unit by asking them to make a "fake" rock. Now have each student display and discuss the results of his or her attempt to make a rock. Emphasize the connection between what the students did and the processes involved in the natural world. Emphasize the variations in the procedures and point out the tremendous variation in nature. Make a summary chart of the students' efforts and the processes of nature.

Right Mode:
Organize a field trip to find real rocks and fake rocks in the community. The trip will have the students visit sites in nature and in the community in which they live. Many storefronts, churches, and graveyards are rich sources of a variety of rock types. The lobbies of banks, schools, and businesses are also abundant with natural stone.

And even more common will be the vast array of "fake" rocks. Concrete, brick, asphalt, and macadam are all examples. The terrazzo and formica tops of counters and window sills also provide a wealth of "fake" rocks.

To culminate the unit ask the students to explore which rocks (real or fake) come from the community. Which create the most impact on their environment? What kinds of things can be done to lessen and to enhance that impact?

COMMENTARY ON QUADRANT FOUR
Quadrant Four is a return to the realm of Concrete Experience. This return is not a petulant dash away from Abstract Concepts, but rather a mature natural shift which brings a quality of reason and stability to one's relationship with the concept. Perhaps it is this move to Quadrant Four at the end of a learning cycle that provides the "enlightened subjectivity" Margaret Mead envisioned.

This return to Concrete Experience honors the notion that education was born to serve life. The narrow, historical perspective of education has held that life was born to serve schooling. This is the tacit meaning given to students who graduate feeling they have "escaped" a form of incarceration in an inside-out system. **4MAT** is designed to restore much of the original integrity of the idea that education must serve life.

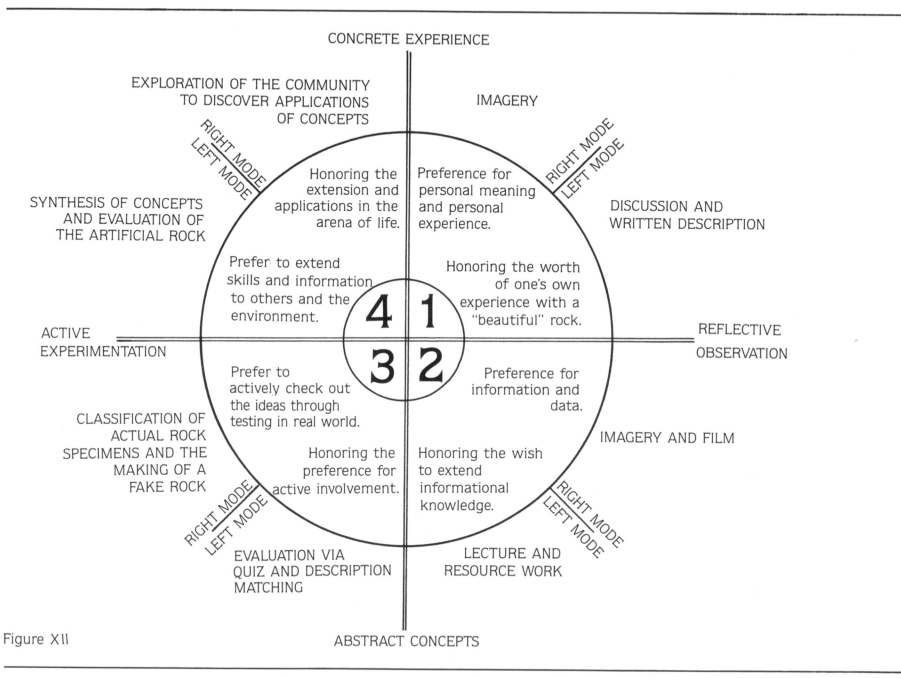

CONCRETE EXPERIENCE

EXPLORATION OF THE COMMUNITY
TO DISCOVER APPLICATIONS
OF CONCEPTS

IMAGERY

RIGHT MODE
LEFT MODE

SYNTHESIS OF CONCEPTS
AND EVALUATION OF
THE ARTIFICIAL ROCK

RIGHT MODE
LEFT MODE

DISCUSSION AND
WRITTEN DESCRIPTION

Honoring the
extension and
applications in the
arena of life.

Preference for
personal meaning
and personal
experience.

Prefer to extend
skills and information
to others and the
environment.

Honoring the worth
of one's own
experience with a
"beautiful" rock.

ACTIVE
EXPERIMENTATION

4 1
3 2

REFLECTIVE
OBSERVATION

Prefer to
actively check out
the ideas through
testing in real world.

Preference for
information and
data.

CLASSIFICATION OF
ACTUAL ROCK
SPECIMENS AND THE
MAKING OF A
FAKE ROCK

IMAGERY AND FILM

Honoring the
preference for
active involvement.

Honoring the wish
to extend
informational
knowledge.

RIGHT MODE
LEFT MODE

RIGHT MODE
LEFT MODE

EVALUATION VIA
QUIZ AND DESCRIPTION
MATCHING

LECTURE AND
RESOURCE WORK

ABSTRACT CONCEPTS

Figure XII

The 4MAT System progression
provides experience
for all students
in each learning style quadrant.

Left and right modes are honored
through the use of both types of activities
as well as
the learning modalities,
auditory, visual, kinesthetic, and symbolic/abstract.

The 4MAT System
intentionally addresses
the learning style differences of all students,
while honoring a whole-brained approach to instruction.

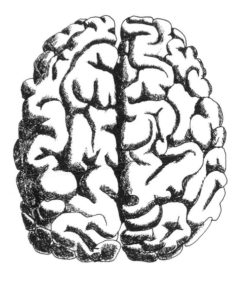

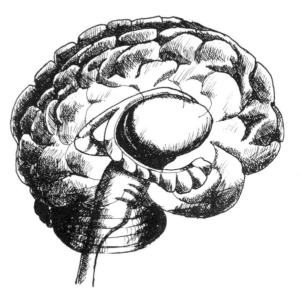

Multi-Modality Learning

In the ROCK lesson unit, we were intentionally conservative in regard to Learning Modalities since many science teachers do not use instructional techniques involving high kinesthetic or auditory expression. For example, the culminating activity in the ROCK unit could have required students to:

write a play with music expressing the ways that rocks form, or

create a skit about classification characteristics, or

create rock origin games to teach other students.

After visiting their community exploring for real and fake rocks, the students could create music, dance, and art that defended the use of real or fake rocks. The whole drama could be a multi-media debate about environmental impact.

Remember the approaches we are offering here are based on studies from a variety of sources. Scholastic achievement has been affected positively by intentionally using movement, music, and art to stimulate more whole-brained involvement in learning. Further, scholastic performance has been raised by increasing flexibility and fluency in thinking processes.

In the lesson units that follow in Part II, notice the variety of ways we suggest Learning Styles, Learning Modalities, and left and right brain information processes are called into the service of a holistic perspective in learning.

How To Write Your Own 4MAT Units

Things to keep in mind

The most critical first step in writing any teaching unit is a clear understanding of the concept to be taught. Frequently, we find teachers do not have the concepts they are teaching adequately defined in their own minds. Our experience has shown it is often not easy to reach to the heart of a concept's underlying simpler structure. Yet teachers must do this well if we are to help our students grasp the wholeness of concepts. The professional judgement of the teacher is an important factor in deciding what concepts are meaningful to teach. That judgement cannot be limited to the rote response that "It's in the text," or "It's in the curriculum book."

Often we are asked to define the term "concept." A concept is simply an invention of the human mind that attempts to simplify, to structure, and to order a complex, interactive relationship in our environment by describing it with words. The authors have synthesized the definition by Novak and Gowin[12] and Gregory Bateson. Our version is as follows.

Concept: a relationship between events, objects and processes.

This working definition helps teachers to focus on the learner's interactions with the world, their perceptions of the patterns formed by the interaction of objects, events, and processes. These interactive encounters are the foundation of the understanding of concepts.

Learning starts far earlier than we usually acknowledge. We now know that infants in the womb begin to acquire the pattern or song of their parents' language. They will know and respond to their mother's voice immediately after birth. They also dream in utero and demonstrate many other capacities for sorting out regularities in the patterns of objects and events in their sheltered environment. Before the age of three, most children have already acquired the basic regularities of speech. Mother, father, sister, brother, flower, dog, cat, colors, birthday, lunch time, nap time, up, and down. Many relationships between objects, events, and processes have been generally mastered. As children learn to acquire language rules and additional labels for objects and events, their communication of pattern recognition is also expanded. Young children are natural learners because they are committed to the process with their whole body and mind. Before they begin school they have seen elaborate patterns of regularities and have internalized these in their own personal way.

Schools often teach skills and facts in boxes. And somehow these skills and facts have become ends in themselves, isolated entities, separated from their larger meaning. Without meaning, there can never be understanding, and here we mean both personal and societal meaning. Meaning is the core idea formed by mentally combining all the characteristics and particulars into a useful construct.

An inherent problem in science education is that very often young children internalize what Mary Budd Rowe[13] calls "naive concepts." These are misunderstandings of science. Gravity provides a prime example. Children, and most adults, believe that heavy objects will fall to earth faster than lighter ones. Even when confronted with a demonstration of the behavior of falling objects in a vacuum, they still believe that a nail made of lead will fall faster than one made of aluminum.

Jerome Bruner[14] indicates significant learning occurs when the learner encounters discrepant events in his or her perception of the patterns of objects, events, and processes. This has been our experience as well. It often takes a shock of surprise to disrupt an entrenched "naive concept."

It is this encounter with discrepant events that generates the magic of science. It is this curious "Aha!" that makes Mr. Wizard popular. It is this encounter that makes scientists engage in unique quests to discover new dimensions in our constructs and world views.

It is difficult to let go of naive concepts. There must be personal hands-on verification if incomplete and inaccurate notions are to be superceded. The exercises we recommend in Quadrants Three and Four give the hands-on verification that is crucial if learning is to take place. Quadrant Three allows students to explore a concept in the full-being way they did as young children. Quadrant Four allows students to engage in concrete experience which extends the concepts into their lives.

Concept: a relationship between events, objects and processes.

START WITH A CONCEPT

INSURE THAT THE CYCLE ESTABLISHES PERSONAL MEANING

CREATE A CYCLE OF ORGANIZED EXPERIENCES AROUND THE CONCEPT

PROVIDE ACCURATE AND RELEVANT INFORMATION

ALLOW FOR ACTIVE HANDS-ON VERIFICATION

ENCOURAGE APPLICATION OF THE CONCEPT IN LIFE

THEN TEACH THE CYCLE HONORING THE CONCEPT, THE STUDENT AND YOURSELF

Key Things To Consider

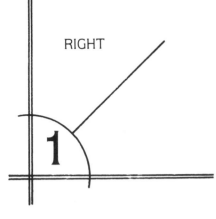

RIGHT

1

Step One

Immersing in an experience
Quadrant One, Right Mode Learner most comfortable
Teacher Role: Motivator
Method: Discussion
Question to be answered: Why?
"Help them establish a reason."

Step One

In Quadrant One we strive to clarify the reason behind the learning. We mutually address the question "Why?"

The Right Mode of Quadrant One is committed to creating a concrete experience related to the concept. Imbue the experience with meaning so students are able to see connections within their own experience.

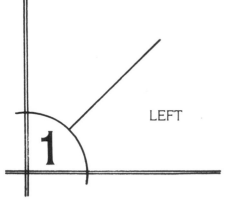

LEFT

1

Step Two

Reflecting on experience
Quadrant One, Left Mode Learner most comfortable
Teacher Role: Witness
Method: Discussion
Question to be answered: Why?
"Help them clarify reasons."

Step Two

The Left Mode aspect of reflecting on experience lies in the quality of analysis. Now the students examine the experience. The method is discussion, which is the method in the first quadrant, but the focus has changed. The students are now asked to step outside the experience and look at its parts.

Quadrant One: Right and Left Modes

You must provide a common experience for the learners that becomes a basis for sharing and discussion. Often this can be done as a homework assignment, for example, asking the students to bring in a "special" rock. Then you can use the rock as a vehicle for awareness, perception, and student interaction.

You should encourage students to identify and express personal feeling, values, and beliefs relative to the concept through simulations, discussion worksheets, checklists, rating scales, or in structured discussions.

You should provide a means for connecting the reason for learning the concept to the relevant relationship it has to their lives.

You should summarize and review similarities and differences in students' perceptions, data beliefs, and values to establish clearly the range of diversity in what is known, and the many ways in which it is known.

You should provide both the mechanisms and a multimodal forum for students to share what they know and believe about the concept.

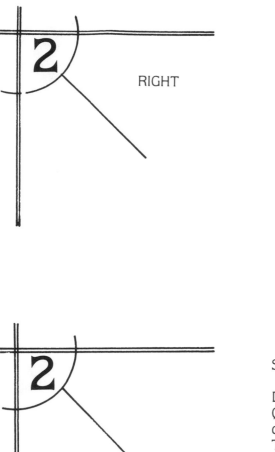

RIGHT

LEFT

Step Three

Integrating observations into concepts
Quadrant Two, Right Mode Learner most comfortable
Teacher Role: Teacher
Method: Informational
Question to be answered: What?
"Teach it to them."

Step Four

Developing theories and concepts
Quadrant Two, Left Mode Learner most comfortable
Teacher Role: Teacher
Method: Informational
Question to be answered: What?
"Teach it to them."

Step Three

The Right Mode Step of Quadrant Two attempts to deepen reflection, with the goal of formalizing and ordering of the concept.

When you design Step Three, Quadrant Two, the Right Mode Step, look for another medium, another way of looking at something that engages the senses while simultaneously affording the opportunity for more reflection. Remember you are moving the students from the concrete to the abstract, and Reflective Observation is the gateway. You want to create an activity that causes them to mull over the experience and analysis just completed in Quadrant One, and assists them in formulating and deepening their understanding of the concept, the purpose of Quadrant Two.

Step Four

The Left Mode Step of Quadrant Two takes your students to the heart of conceptual information.

It is in this step that the formality of the way the concept organizes experience is validated. Here is where the learner is provided with the information related to the concept so as to understand in conventional ways. We are not interested in rote memory, the antithesis of thinking. We are stressing information that relates to the core of the concept.

Quadrant Two: Right and Left Modes

You should recognize that two key strategies for input here are first, to establish a learning experience that provides students with a gestalt or perspective of the concept being taught. And secondly, to deliver the specifics of cognition, content, vocabulary, and conceptual relationships specified in the school system's curriculum or course of study.

You must offer effective right mode activities that can lead learners to larger perspectives of the concept patterns. Among these are guided imagery, overview films, concept mapping (see page 163), and structured multisensory experiences such as creative dramatics (see page 145), or art projects (see page 146). These kinds of activities lead students to new connections to the concept patterning that initiates Quadrant Two.

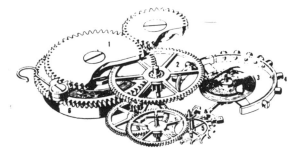

You can serve the left mode processing through lectures, text assignments, reference reports, guest speakers, audio and visual presentations, library searches, and computer assisted instructional programs. All are fundamental Quadrant Two delivery systems for providing students direct access to the specific concept related information you need the students to know.

Note that these last two steps, Steps Four (Quadrant Two) and Five (Quadrant Three), are Left Mode techniques. Step Four appeals to the Analytic Left Mode learners, and Step Five appeals to the Common Sense Left Mode learners. One can easily see the value of these two steps for all learners, but the exclusive teaching in only these two modes make it extremely difficult for any other learner to succeed.

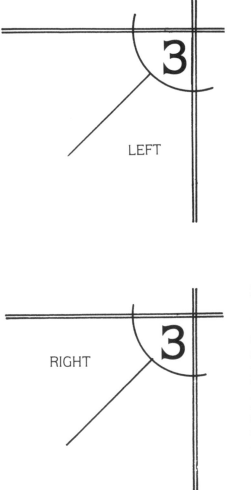

LEFT

RIGHT

Step Five

Working on defined concepts
(Reinforcement and Manipulation)
Quadrant Three, Left Mode Learner
most comfortable
Teacher Role: Coach
Method: Facilitation
Question to be answered: How does this work?
"Let them try it."

Step Five

In Step Five, a Left Mode approach, the students react to the givens. They do worksheets, use workbooks, etc. These materials are used to reinforce the concepts and skills taught in Quadrant Two.

Step Six

"Messing around"
(Adding something of themselves)
Quadrant Three, Right Mode Learner
most comfortable
Teacher Role: Coach
Method: Facilitation
Question to be answered: How does this work?
"Let them try it."

Step Six

Step Six is active thinking. This is learning by doing, and its essence is problem solving.

Quadrant Three: Left and Right Modes

You must remember this is the Quadrant where students try out the concepts they have learned. It is critical that they try out what they have learned in the real world. It is here that we set up controlled conditions where students can test their understandings.

You must afford the students the opportunity to demonstrate mastery of the skills, labels, and basic operation of the events and objects they have experienced in the more abstract terms of Quadrant Two. Lab workbooks that accompany texts with their "label the parts," "fill in the blanks," and "draw the apparatus" assignments provide the needed link for common sense learners. These verify student mastery of the labels of objects and events that are associated with the regularities of a known conceptual pattern.

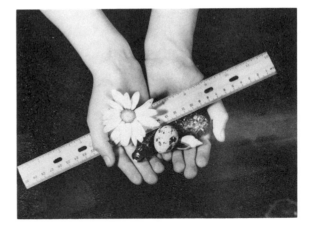

For Right Mode fulfillment, you must allow the students to explore, tinker, and get the concept in their hands. The heart of a quality science program is in Quadrant Three, Right Mode. This is where students are challenged to design their own open-ended explorations of the concept.

Laboratory experiments, field studies, and environment explorations where students are involved in the design of key components are very effective mechanisms for stimulating student discovery and first hand encounters.

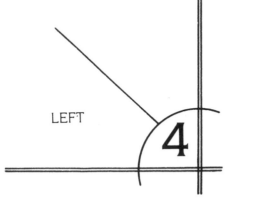

LEFT

Step Seven

Evaluating and synthesizing for usefulness or application
Quadrant Four, Left Mode Learner most comfortable
Teacher Role: Evaluator/Remediator
Method: Self-Discovery
Question to be answered: If all this fits together, what does it mean?
"Let them teach it to themselves and to someone else."

Step Seven

This is the step where the students are asked to analyze what they have planned as their "proof" of learning.

Here the students are required to organize and synthesize what they have learned in some personal, meaningful way. This is the "grand finale" of their Left Mode journey in this concept unit. As this step closes, they should have a cumulative sense of the exploration through all the previous steps.

RIGHT

Step Eight

Doing it themselves and sharing what they do with others
Quadrant Four, Right Mode Learner most comfortable
Teacher Role: Evaluator/Remediator
Method: Self-Discovery
Questions to be answered: How can I apply this? What can this become?
"Let them teach it to themselves and to someone else."

Step Eight

The last step of the lesson unit, Step Eight, is when the students share what they have learned with each other.

They are encouraged to take the responsibility for making their own sense of what they have learned by applying it in life. The culmination of this step is to extend one's sense of having learned the concept to the beginnings of living the gifts of that learning.

Quadrant Four: Left and Right Modes

You provide benefit of your expertise to assist the students in synthesizing the results of their learning. This may be done in a variety of ways, but the focus is on the integrity of understanding the concept being studied.

Peer teaching by investigative teams, committee reports, graphic mural presentations, and student demonstrations are a few ways students can share information and establish what they have learned. Here the teacher facilitates by providing meaningful feedback. The key strategy is helping students focus on what they have experienced and its usefulness and relevance to their lives, their community, and/or their world.

The final culminating activity in Quadrant Four should challenge the creativity of students to find new ways of sharing what they have learned with others or to invent new applications and future uses for the mastered material.

The cycle spirals on as creative applications stimulate the value and belief system of the learners to expand, grow, and engage new concepts to explore. And the cycle begins anew moving back into the Quadrant One, Right Mode, "I know, I believe."

And on it goes.

THE LEARNING CYCLE IS A
COMMITMENT TO BRINGING THE
STUDENT'S MIND INTO HARMONY
WITH THE WAY THE WORLD WORKS.

The 4MAT Rhythm

One of the important characteristics of **4MAT** as an instructional technology is its rhythm. The rhythm assures diverse roles for the teacher as he or she moves around the circle. At the same time **4MAT** requires that all learners be stretched into all four quadrants. This movement through the **4MAT** Cycle assures that all students can be winners at least some of the time as their learning preferences are favored.

The **4MAT** Rhythm begins with a wide open right mode interaction with a concept and the honoring of students' feelings, attitudes, and past experiences concerning that concept. These feelings, attitudes, and experiences that are part of the students' lives are then focused down to a discussion of key common elements as the teacher facilitates a sharing and discussion of these experiences. Then **4MAT** Quadrant Two opens out again in order to explore the right mode big view. This is followed by analysis, convergence, and focus as the teacher leads the students through the known facts, theories, etc.

As one moves into **4MAT** Quadrant Three, the cycle shifts. In Quadrants One and Two, the teacher was the main dominant force. Now "Doing" takes precedence and the students become the main actors. The first student act in Quadrant Three is to focus and analytically assess his or her own understanding of the concept via worksheets, quizes, tests, or other standard devices. The second student dominated part of Quadrant Three has the students testing the concept in action. This is far more divergent but it is still wedded to the basic fabric of the concept.

In **4MAT** Quadrant Four, the first act is to review, synthesize, and blend together the various explorations into an intact understanding. Although this step is one of synthesis, it is true to analysis, detail, and convergence. In the final step of the cycle, the students are again urged outward. They try to apply their accumulated experience to the world at large.

Often the results of the rhythmic journey are so motivating that the cycle can begin again. The students can again move to the realm of Quadrant One's commitment to personal meaning and continue the quest.

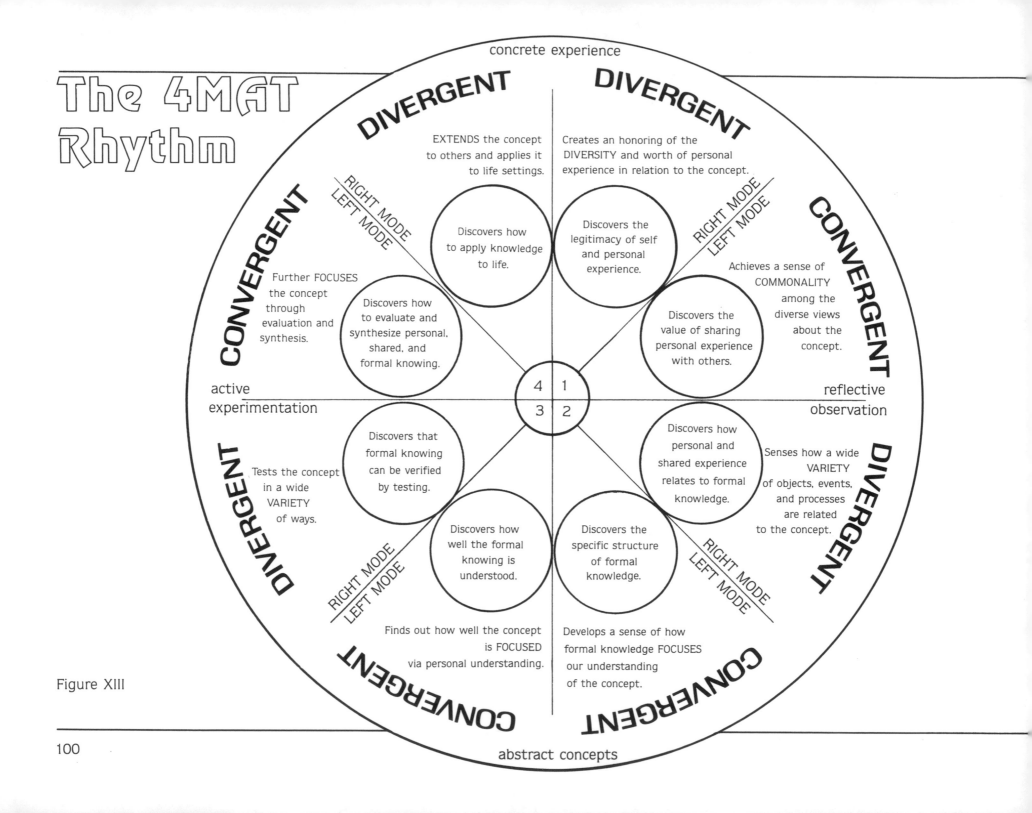

The 4MAT Rhythm

concrete experience

DIVERGENT | DIVERGENT

EXTENDS the concept to others and applies it to life settings.

Creates an honoring of the DIVERSITY and worth of personal experience in relation to the concept.

RIGHT MODE / LEFT MODE

CONVERGENT

Further FOCUSES the concept through evaluation and synthesis.

Discovers how to apply knowledge to life.

Discovers the legitimacy of self and personal experience.

RIGHT MODE / LEFT MODE

CONVERGENT

Achieves a sense of COMMONALITY among the diverse views about the concept.

Discovers how to evaluate and synthesize personal, shared, and formal knowing.

Discovers the value of sharing personal experience with others.

active experimentation

4 | 1

3 | 2

reflective observation

DIVERGENT

Tests the concept in a wide VARIETY of ways.

Discovers that formal knowing can be verified by testing.

Discovers how personal and shared experience relates to formal knowledge.

Senses how a wide VARIETY of objects, events, and processes are related to the concept.

DIVERGENT

RIGHT MODE / LEFT MODE

Discovers how well the formal knowing is understood.

Discovers the specific structure of formal knowledge.

RIGHT MODE / LEFT MODE

CONVERGENT

Finds out how well the concept is FOCUSED via personal understanding.

Develops a sense of how formal knowledge FOCUSES our understanding of the concept.

CONVERGENT

abstract concepts

Figure XIII

100

The Right Activity In The Right Quadrant

Many new **4MAT** users struggle with the question, "How do I know if I have the right activity in the right quadrant?"

We suggest that the rhythm patterns just described and the type of response teachers want to stimulate in students are the best guides.

The appropriateness of input is revealed in the output it elicits in each student. Teachers must be alert to varying input and output.

Quadrant One/Right Mode
Learners should have diverse responses based on personal knowing, life experiences, and expression of beliefs and values.

Quadrant One/Left Mode
Assignments should stimulate common insight convergence, analysis of similarities and differences through discussion and interaction.

Quadrant Two/Right Mode
Assignments should enrich students' divergent insights into the concept by stretching into new perceptions of the whole, the patterns, the gestalt in order that the students may build new connections.

Quadrant Two/Left Mode
Activities should build specific meaning, common understanding convergence, and lead to further depth of understanding of related objects and events.

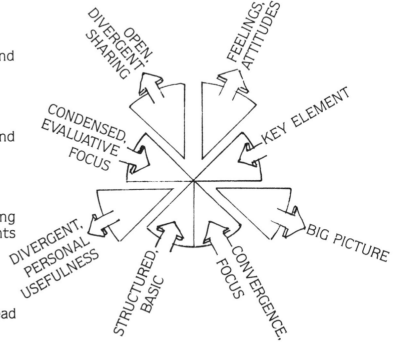

Quadrant Three/Left Mode
Converge student understanding through demonstration of competency with labels and constructs relevant to the concept. The students must master the "basics" of the concept.

Quadrant Three/Right Mode
Elicit diverging student responses as students add something of themselves by trying out the concept in real world applications.

Quadrant Four/Left Mode
Choose activities that converge students' knowing through participatory sharing to common insights and experience.

Quadrant Four/Right Mode
Stimulate widely divergent creative application of the materials learned.

Balancing The Cycles In Learning Science

"It's time to change the baby's bath water, but please don't throw out the baby."

In the last few years, all three of us have traveled and worked extensively with teachers. Everywhere we go we find people who sense we must improve science instruction to meet the demands of a whole new human era of knowing. Yet educators frequently persist in the notion of dropping what we have been doing for years and adopting something totally new.

We do not believe this is an intelligent course of action. We believe we can blend and mix the good practices of the past with insights **4MAT** has to offer. Consider the following:

We find that the Quadrant Two biased textbook-based programs are excellent for some students. If labs are used in them, it is most often validation of text delivered concepts rather than labs which stress discovery.

We find the lab programs and alphabet-process programs which favor Quadrant Three students are excellent for about one-fourth of the students all the time. They are not as effective for the other three quarters. Yet they differ from standard text programs in one significant way. . .the labs stress discovery. Recent studies[15] show that the hands-on approaches of the science-process programs resulted in a large body of students who "liked science." They showed this by choosing science as their first or second favorite subject. The same studies confirm that the students' scientific knowledge was improved by these same programs.

The greatest single difficulty with the science process approaches is that their delivery is anchored firmly in Quadrant Three. Less than 25% of our elementary and secondary teachers have a personal preference for Quadrant Three themselves. More than 60% of the teachers' personal learning style preferences are in Quadrants One and Four. Is it any wonder that we have encountered resistance in "turning on" elementary and secondary teachers to process science programs that are often the antithesis of the way they personally prefer to learn?

Text-based programs fare better in implementation, even though again they are not in the dominant learning style preference patterns of most teachers. Textbooks are the cornerstone of instruction in our culture. And research confirms teachers overwhelmingly use this approach. Remember text approaches are dominantly Quadrant Two. In the past many science teachers experienced training programs where participants read about and were told how to do process-inquiry instructional strategies. These are favored by Quadrant Three. Without experience these teachers were then expected to implement process science programs. Sometimes they were even evaluated on how well they did.

This process is doomed to failure. One cannot expect teachers to change their instructional delivery systems unless they are exposed to training in which a changed delivery system is part of the process they undergo. We are continually amazed to attend conferences where "Creative Problem-Solving" sessions are lectures read from podiums in droning voices. If teachers are to partake of the benefits of the new learning styles and brain/mind research, they must be afforded the opportunity to master these insights in a holistic training program that itself honors diverse learning styles and uses multiple methods of instruction.

There is an additional work burden in preparing lab activity. Teachers also need adequate supplies and equipment if lab techniques are to afford experimental and exploratory activities. Contrast this with the convenience of a simple, concrete science text with all the questions and key answers neatly packaged, guide question for each chapter, and a limited number of "safe" demonstrations to liven things up, and we readily see that the teachers' lessons simply become "the chapter and pages we will cover today. . ." When the door closes in a science classroom, the balance tilts strongly to "TELL-TEXT-TEST" science teaching.

We believe this is just not good enough!

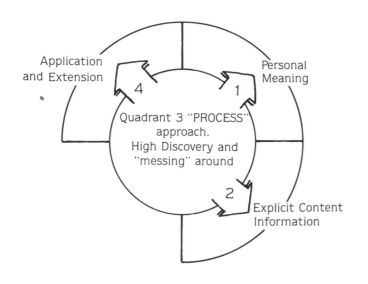

Application
and Extension

Personal
Meaning

4

1

Quadrant 3 "PROCESS"
approach.
High Discovery and
"messing" around

2

Explicit Content
Information

We believe, from our own first-hand experience, that when you put balance in instructional delivery back into a science program, you have excited teachers and students.

In our workshop we explore the bias or preference of the following programs: Elementary Science Study (ESS), Science, A Process Approach (SAPA), Science Curriculum Improvement Study (SCIS, SCIS II), Science 5/13 (McDonald/Nuffield), Individualized Science Instructional System (ISIS), Biological Sciences Curriculum Study (BSCS - all versions), Introductory Physical Science (IPS), Earth Science Curriculum Project (ESCP), Chem Study (CS), Physical Sciences Study Committee (PSSC). . .all of which are primarily based within the preferences of Quadrant Three with secondary commitment to Quadrant Two. Only minor excursions are usually made into Quadrants One and Four.

Process approach science materials can be rounded out into a learning cycle by being attentive to the other Quadrant characteristics. Most teachers quickly find this to be a fairly simple and even joyful process. The Quadrant One assignment is often given as homework, which leads to group discussion in class. It is focused on connecting the concept to personal meaning; Quadrant Two is delivery of information (films, lectures, text readings, computer programs); and Quadrant Four is the extension application which also can often be accomplished as homework or in class time. Any teacher who already possesses the teaching gifts of Quadrant Three can easily make science "magic" by adding the dimensions of "Oneness, Twoness, and Fourness" that 4MAT encompasses. (See the sample lesson units in Part II.)

When judging the science programs of the past two decades, it is clear that they were remarkable improvements in traditional materials. Science programs such as BSCS demonstrated considerable balance in their original creation and delivery. These programs combined text readings with process labs, with films, and a variety of handbook excursions that provided a desirable balance in instructional delivery. However, there were noticeable gaps in the treatment of Quadrants One and Four. The problem was that not enough teachers were trained to use the total program as it was conceived, and ultimately we ended up with a process textbook inadequately used as an information text, a role it was not created to fulfill. The result has been a shift back to the **Modern Biology** [16] text which has been upgraded with new pictures and some BSCS-like concessions in format. The text is easier to use as the built-in text/lab connections are not as interdependent as the BSCS versions.

In contrast, the science textbooks of today have been vastly improved in format. Great effort has been made to stimulate visual modality with color, pictures, diagrams, and charts. The texts are very readable and provide a detailed, instructional, cookbook guide for teachers including how to do the key demonstrations which are illustrated completely for students.

One of the better examples of a conceptual-based science text that has within it many key elements for text accessible to a greater variety of learners is **Life A Biological Science** by Brandwein, Yasso, and Brovley[17]. In their approach, they provide access to traditional biological concepts, connect related concepts, review the key concepts, provide concept-based self tests, provide questions of choice that relate value applications in the "real world," provide suggested extensions of the concepts for further investigation. . .all delivered in an inquiry questioning style.

However, even a text such as this lacks balance in presentation. A majority of the text space focuses on delivery of information and labels about objects and events (Quadrant Two, Left Mode). However, when this text becomes an integral part of the **4MAT** Wheel lesson unit, the Quadrant One introductions provide a stimulus for students to want to find out more using the text as a prime information source, (a vehicle for Quadrant Three, Left Mode, and Quadrant Four, Left Mode). This leaves the teacher and students the relatively simple task of applying their own creative insights in generating the Quadrant Three and Four Right Mode experiences.

4MAT makes existing texts central to patterning the instructional cycles without having them become restrictive and dominating.

In summary, most of the commercial science instruction materials are excellent for providing students and teachers meaningful science instruction, but in a very narrow range of learning preferences. Most are heavily focused in either Quadrant Two or Three, or some in both. Few provide significant learning challenges for Quadrant One and Four preferences.

4MAT can play a fundamental role in providing balance to any science program by clearly identifying in a systematic way where and how resources traditionally used fit into a holistic science instructional program, and where teachers and students must generate complimentary instructional activities and experiences to assure all students are winners at least some of the time.

The result is the added dimension of teachers actively engaged themselves in a creative inquiry process. Teachers design instruction that models what they are teaching.

Science is more a process of creative invention than discovery. It's that sense of invention that is most missing in schools.

Learning Styles and Laboratory Experience

During his years as a teacher Bill Hammond noticed a distinct set of student behaviors at work in his science classrooms. Although this was long before **The 4MAT System** was written, there are remarkable connections. The students we now designate as Quadrant One, Quadrant Two, Quadrant Three, and Quadrant Four learners, worked very differently in lab situations.

The Quadrant One Learners, those most comfortable in Quadrant One activities, would spend inordinate amounts of time deciding whether or not they "felt" like doing the lab. Once committed they would wander about engaging in discussions and trying to extract what the lab meant from other students. All the while they would lament that they could not see why they were being asked to do this.

In short they did not like doing labs and gained what they could in lab activities by discussing the topic with others.

The Quadrant Two Learners, those most comfortable in Quadrant Two activities, usually had read a chapter ahead and knew the lab objectives. If lab forms were handed out prior to the lab, then the Two's would have them filled out before they came to class. They would go through the motions in the lab, but the results had already been accurately fabricated. Another favorite trick of the Two's was to seek out a Quadrant Three Learner as a lab partner.

In summary, Quadrant Two Learners do not like labs. However, they consider them necessary and do value the data that come from them.

The Quadrant Three Learners,those most comfortable in Quadrant Three, are lab students. Labs were invented for them. They cherish the opportunity to test the reality and applicability of abstract concepts usually dealt with in lectures and books. They are precise and thorough and their lab work stands as a model for all.

In summary, the Quadrant Three Learners love labs.

The Quadrant Four Learners, those most comfortable in Quadrant Four activities, are anathema in labs. They set up the apparatus and then begin a migration that takes them to nearly every point in the lab. If the teacher is accomodating, the Fours will wander about with unsolicited comments and suggestions to all working students. They have recommendations for optional experiments and advice about how things could be better. Yet in the midst of this apparent abandonment of purpose, they amass a wealth of data and gain remarkable insights into dozens of apparently unrelated topics. They know who is dating whom—who is "cheating"—whose data are the most valuable and where new discoveries have been made. They will almost always find a way to report for the group.

In summary, the Fours tend to treat the lab as a field trip. Once the field study is done, they are perfectly willing to report for the group.

All of this suggests the possibility of student laboratory "teams." Research in cooperative learning has suggested much can be done to enhance success. We do not need to specifically type or label students, but rather using approaches such as **4MAT** we can call forth explicit differences in styles of learning and thus enhance the possibility for the natural formation of heterogeneous teams. Something needs to be done, particularly in science education where tradition has set the pattern TELL-TEXT-LAB-TEST.

This pattern favors the **4MAT** Quadrants in this order: Two-Two-Three-Two. It is simply not enough.

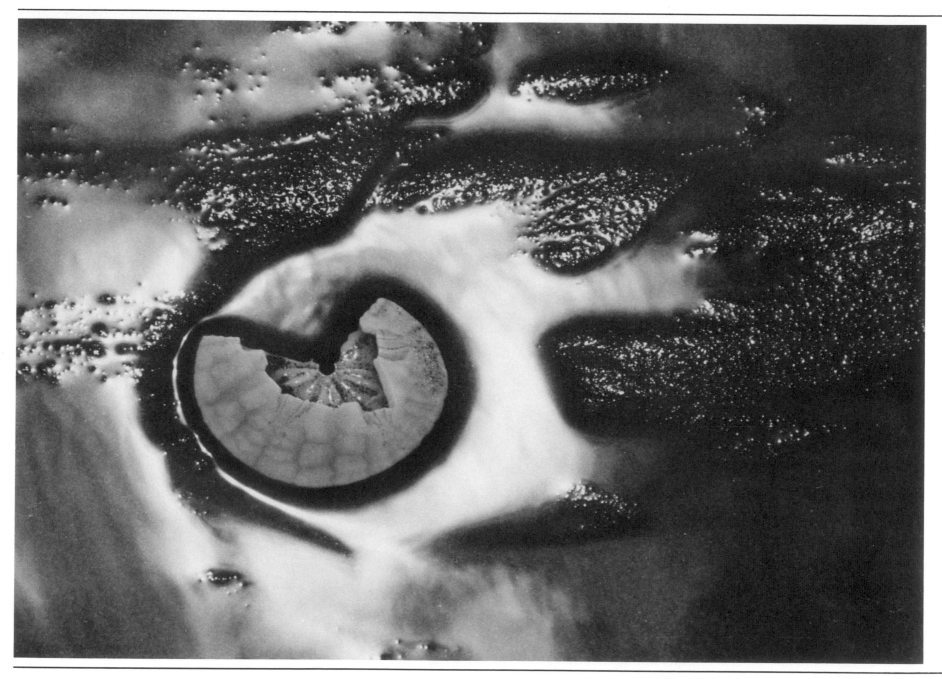

4MAT And Science

One of the games information statisticians frequently play is to proclaim that the next ten years will witness the creation of more knowledge than in the entire history of humankind. We are also told that ninety percent of all scientists who ever lived are alive now. We can extrapolate freely in the face of such claims that any ten year old American child with access to a computer has more data availability than Leonardo or Galileo.

We know that within ten years there will be more information processers in American homes than television sets at the time President Kennedy was assassinated. We also know the networks of information access will be thousands of times more elaborate than they are now. Schools will shift from computer education programs to programs of computer use. The entire contents of the Library of Congress will be as close as a phone and a modem.

In the face of all this, **lifelong learning** will be as natural an option as good nutrition and the evening news. Approaches such as **4MAT** tend to nurture a wider view of learning. **4MAT** emphasizes that learning is personal and bonded to the meaning we choose to give to life. Learning is also information and data. Within that realm, learning has an integrity of its own. Accuracy, precision, and adherence to the patterns of inquiry are a part of wisdom. Learning is also action. Things learned give us something to do. We can take all manner of forms of action on our lives because of what we have learned. Learning is also the opening that gives us access to wider realms of the human spirit. We can learn to teach and to attempt to heal the insults and disparities suffered by humans and the planet upon which we live.

Science is a way of knowing that is a natural quest of the human mind. It is as inherently a part of life as the arts once were. It remains for us to see and understand it that way. Science could be an entertainment for children in the crib and an adventure for children in school. It could be a richly rewarding companion for the adult mind.

Science can be restored to the wholeness that characterizes the way true scientists work. We can restore this wholeness without the loss of integrity. On the contrary, the real integrity of science lies not in science as a body of knowledge, but science as an expression of the human spirit. When the human is returned to the equation, science can be experienced as a quest. It will take the realization that objectivity is a noble goal but a shadowy reality. The words of Margaret Mead again will become poignant—"We must give up the notion that we operate within the realm of objectivity and instead attempt to establish an enlightened subjectivity."[18]

It is this admission of an enlightened subjectivity that returns science again to the center of humanness. It restores to it the spiritual—the mythic and the most noble qualities that have guided humans on their immense journey. Gone will be the substance of the prejudices against science which created the image of the white-coated automatons who willingly ignore ethics and morality.

Science is not, nor has it ever been, a narrow slice of the exploration of mind. The commitment to accuracy and formal structure are part of the art of science. Unfortunately for some, the case was not presented as an art form, but rather as a prison of conformity in which the freedom of the mind was confined. Fortunately this image is waning. This book is an effort to restore wholeness to the human explorations we call science and to establish a context of wholeness within the realms of science instruction and learning.

4MAT And Self

4MAT has another facet which is vital to the learning cycle. This has to do with the image of self that is such a central part of lifelong learning. Each learner who is guided through the cycles of experience inherent in a basic **4MAT** lesson unit meets many faces of knowing.

Quadrant One Learners find in the activities that personal meaning and personal experiences have real validity. They discover, that regardless of the concept being studied, there are experiences in one's own past that are linked to the ideas under study. Indeed the linking of ideas to the personal experiences of the student is the hallmark of the master teacher. Each of us is a reservoir of tacit experiences that, when formalized, afford us insights from the reality of our personal histories. The teacher who gives us the confidence to be our own primary source gives us the gift of knowing that much of wisdom is innately our own.

In Quadrant Two the self is enriched with the power of accumulated knowledge. The learner is strengthened by the formal processes and information of the experts past and present. If these concepts are presented with the integrity and grace of the spirit of inquiry, the learner will find accumulated knowledge an ally and a host for farther journeys.

Quadrant Three allows the self to grow in the confidence that accumulated wisdom can be applied to life. Learners find professions that require "acting upon givens," such as medicine and engineering, require courage. Learning cannot take place without students being allowed to make choices, to explore, to manipulate, to tinker. Nowhere is this more true than in science. Exploration, manipulation, and experimentation cannot be accomplished without students applying what they learn to life in a meaningful way. It is implicit in the very meaning of science that learning be extended beyond the classroom. This can never be accomplished if science instruction is limited to reading another book, filling out workbook pages, or writing another report.

If Quadrant Three provides the courage to guide one's own destiny, then Quadrant Four fulfills the promise. Here the learner is given the excitement of choosing the course of inquiry and guiding it to fulfillment. Here learners discover the true meaning of personal integrity. They develop a way to see how all they have experienced is part of a larger pattern of interconnected human involvement. They learn here they can teach, they can sow the seeds of healing in larger settings. It is here that the self grows to become the inclusive form which weds it to the larger world.

4MAT And Society

Out of the experiences cited above, the individual assumes a remarkably different role as a member of society. By systematically exploring the options of mind in the act of learning, it becomes difficult, if not impossible, to become isolated in any single sector of learning.

Without the emphasis on one style of knowing that continues to plague our schools, we can begin to honor diversity in learners, as well as diversity in knowing. And cycles of instruction can be enormously effective in guiding learners to this kind of diversity.

Without the enforced isolation of a single style of knowing, a rare kind of personal freedom can be attained. And this freedom can transform society.

If each of us is the basic unit of the society in which we live, and if each of us has been rewarded by that society through an education that nurtures wholeness, then we are each the cause and the effect of fulfillment. Perhaps one of the manifestations of that wholeness will be the confidence to restore our basic perception that all forms of knowledge—whether scientific, social, mathematical or even the arts of literature, graphics, sculpture, and dance—are in fact different weavings in the same tapestry of mind. We may come to honor the gifts of our heritage without engaging in the games of fragmentation and compartmentalization which have so grievously marked our past.

The real changes that tomorrow will witness must begin somewhere. Perhaps our own minds and the many ways we can learn are as worthy a beginning point as we can find.

Learning Cycles

The following examples of learning cycles attempt to provide guidelines in several content areas for how specific topics may be explored. Remember that these suggestions were written in the spirit of the unit on rocks presented earlier. None of these has a commentary section. Thus they stand as "bare bones" examples. You and the students must turn them into life.

We offer four kinds of different cycles in the following pages. The first kind is a set of fairly complete learning cycles. These include Clouds, Growing Plants, Number Systems, Mapping, Galaxies, and Sleep.

The second kind is an incomplete learning cycle. In Ancient People, we take you through Quadrant Three, Left Mode. We leave it to you to finish.

In a third kind of representative learning cycle, Cell Growth (hybrid), we paraphrased two assignments from a popular junior high school science text. These assignments each represented a portion of a complete **4MAT** Learning Cycle, so we put them where they best fit and completed the cycle as we saw it. This example shows how we believe standard materials can be adapted to honor whole-minded instruction.

The fourth kind is a "Z" cycle. We want to call it to your attention so you can avoid including it in instruction.

4MAT: Clouds

1 Have the students collect images of clouds for a week. Photos, paintings, drawings are all appropriate. Bring the images to school on a designated day.

Visit T.V. stations or other weather facility. Determine what the basic information is that is needed for forecasting. See if weather broadcasts can be made at school.

Invite the students to write poems, haiku, "wise-sayings," about the images. Create a gallery of images and words. Form discussion groups and have them report reactions.

4 Review and Synthesize the principles of cloud formation, the variety of shapes and the scientific meanings. Link the science with the poetry and images.

2 Re-organize the gallery so it is displayed with the clouds of similar shapes gathered together. Show a film or filmstrip that displays the different cloud types.

3 Have the students set up experiments that explore condensation, evaporation, precipitation, etc.

Have the students label and classify all the images in the gallery. Try to retain the aesthetic appeal of the gallery.

Provide a lesson about the classification of clouds. Emphasize the relation between shapes and how they form.

Growing Plants

4MAT

Visit nurseries and farms. Find out how gardening is carried out in the community. Devise a project that will beautify the school Grounds.

Have the students close their eyes and imagine that they are their favorite plant. Provide a guided imagery that leads them through a life cycle.

Review and synthesize all the previous lessons. Include the poetry, imagery, movement and information. Begin to plan a garden.

Gather in groups of four and discuss how it felt to be a seed, sprout and grow. When they finish have them write a poem about "the needs of seeds"

4 1
3 2

Provide the students with a variety of seeds and nursery supplies. Have them exp-eriment on plant needs for a period of three weeks.

Do A review worksheet on the basic needs of plants.

Lecture on the basic needs of plants. Establish a basic list of conditions to be met for plants to live.

Ask the students to close their eyes and review all the movements that they went through as they "grew." Open their eyes and view a film of time lapse growth.

126

RIGHT MODE | RIGHT MODE

4

Go out into the community and find a number system that is useful to you and one that does not seem useful to you. Create a number system that can help your -ng students.

Explore both the burden and advantage of number systems that apply to you, the other students and the school. Synthesize their interrelatedness.

LEFT MODE

1

Ask the students to imagine themselves as a number. (as large as they want just so they can remember it) Have them close their eyes and imagine being called by that number.

Have the students affix the numbers to themselves. Have them group together in "numbers" that have some similarity. Have them discuss the similarity and report to the class.

LEFT MODE

RIGHT MODE

3

Turn the school into a laboratory and find out which number systems are necessary for the school to function.

Ask the students which number systems that apply to them can be changed without affecting them. Which can be changed without causing confusion.

LEFT MODE

2

Give a lecture on numerical notation systems and how they are used in society. (metrics, measurement, stellar notation etc.)

Have the students explore the number systems they share (phone numbers, address, grade points. etc.) Have them create posters that celebrate all the number systems that apply to them.

RIGHT MODE

LEFT MODE

Mapping

RIGHT MODE **RIGHT MODE**

4 | **1**

3 | **2**

Quadrant 1 — RIGHT MODE
Ask the students to bring in a map that shows all the places they have been. Have them locate their travels on the map. It may be a city state or nation-world map.

Quadrant 1 — LEFT MODE
Gather in groups and discuss your favorite and least favorite trips. Have the students write individual statements about places they would like to go as reported by classmates.

Quadrant 2 — RIGHT MODE
Choose several maps from those brought in by the class. Have the students create a single map that incorporates all the others. (The room may be too small)

Quadrant 2 — LEFT MODE
Present a lecture on the fundamentals of scale and maps. Indicate how information on maps depends on the scale.

Quadrant 3 — LEFT MODE
Ask the students to show which scale should be used to show the relation between the class room and- 1) the office 2) city hall 3) Los Angeles 4) Paris and 5) Venus.

Quadrant 3 — RIGHT MODE
Working in groups of five map the school-ground in detail. First they must establish a scale to use.

Quadrant 4 — LEFT MODE
Discuss the effects of the various scales chosen in mapping the school ground. Synthesize all the earlier activities.

Quadrant 4 — RIGHT MODE
Have each student create a "life-map" that charts the course of their future travels. Help the students determine the "scale" used in describing their lives.

RIGHT MODE RIGHT MODE

Arrange a trip to an observatory or a local star party with telescopes. Look at as many deep sky objects as possible. Have the students write a skit or play to present to younger students.

Invite the students to create paintings of Galaxies and star systems. Use black paper and splatter brush technique ...the images can be enhanced with chalk.

Synthesize the understanding of galactic shapes, sizes and evolution. Focus on the relationship between function, form and aesthetics in galaxies.

Have the students gather in groups of 5 and discuss the images. Have them become a being in one of the systems... discuss life as it exists there.

LEFT MODE

4 1

3 2

LEFT MODE

RIGHT MODE

RIGHT MODE

Have the students construct a model of the galaxy that would fit on a ping-pong table. Have them use turntables, swivel chairs etc. to study swirling systems.

Ask the students to calculate the size of the Milky Way. Have them calculate the dimensions of a scale model with the earth/sun distance = 1 mm.

Lecture about the classification of galaxies. Introduce the notion of galactic evolution. Introduce the idea of astronomic distances.

Using the five basic galactic shapes 1) globular clusters, 2) elipses 3) spirals 4) barred spirals and 5) open spirals. Ask the students to choreograph the class so as to move in ways that the galaxies might.

LEFT MODE LEFT MODE

RIGHT MODE **RIGHT MODE**

Invite the students into the process of dream re-flection (See p. 142). Use the process to estab-lish a "bulletin board" of insights, ideas problem solut-ions gained through dreams.

Invite the students to find out as much as they can about their own sleep habits. Have them interview parents and other rel-atives about how they slept as infants.

Form a panel discuss-ion to explore the effectiveness of the sleep experiments. Relate the results to their earlier sleep histories.

Ask the students to write 2 paragraphs about their sleep habits. In groups of five share the information and determine what is "normal" in sleep habits.

LEFT MODE

LEFT MODE

4 1

3 2

RIGHT MODE

RIGHT MODE

In teams of five have the students combine their lists and develop a way to explore those realms of sleep that they have not previous-ly exper-ienced.

Have the students list the various characteristics of sleep which they have experienced. Establish how the list relates to the states described in the lecture.

Offer the students a comprehensive lecture on sleep. Include dreaming, sleep walking, talking in sleep and the different physiological states. In-clude coma, hiber-nation/estivation.

Create a mural-like record of the variations of sleep patterns of the students. Use large sheets of butcher paper and poster paints. Perhaps show a film on sleep.

LEFT MODE **LEFT MODE**

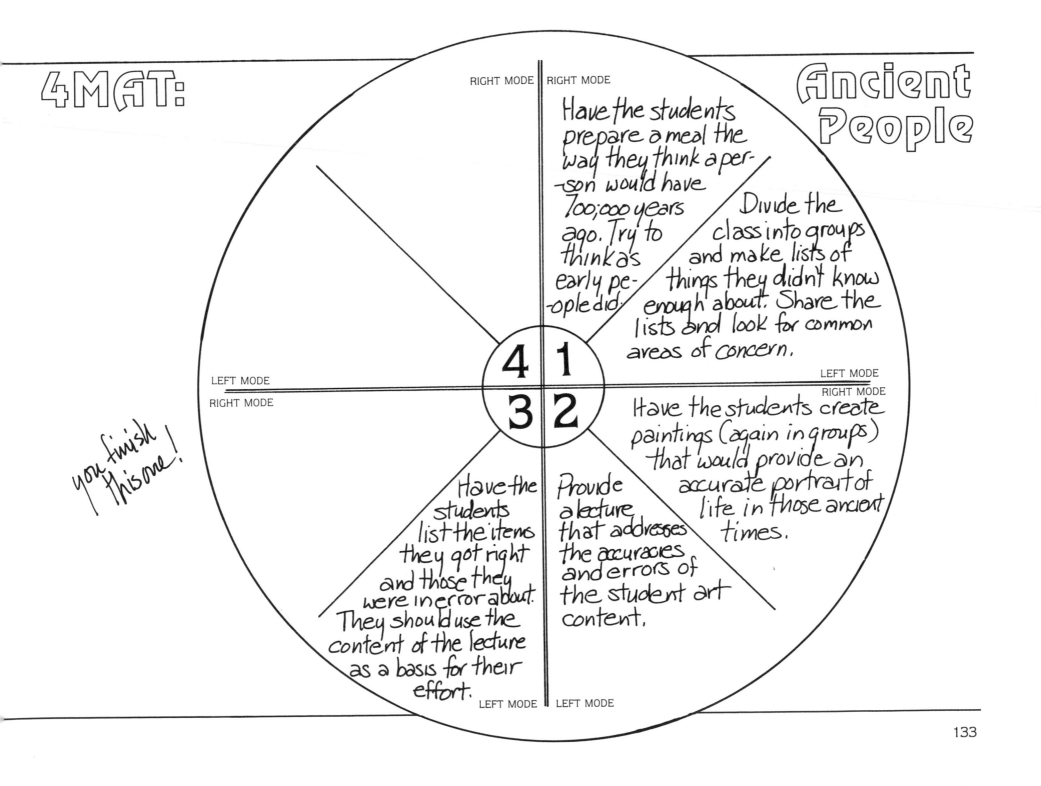

4MAT:

Ancient People

RIGHT MODE | RIGHT MODE

Have the students prepare a meal the way they think a person would have 700,000 years ago. Try to think as early people did.

Divide the class into groups and make lists of things they didn't know enough about. Share the lists and look for common areas of concern.

LEFT MODE

RIGHT MODE

4 1

3 2

Have the students create paintings (again in groups) that would provide an accurate portrait of life in those ancient times.

LEFT MODE

RIGHT MODE

you finish this one!

Have the students list the items they got right and those they were in error about. They should use the content of the lecture as a basis for their effort.

Provide a lecture that addresses the accuracies and errors of the student art content.

LEFT MODE | LEFT MODE

Cell Growth (hybrid)

RIGHT MODE

4 — Visit medical and research facilities in the community. Find out how and why research in controlled cell growth is going on. Determine what cells the students can control the growth of.

Synthesize personal changes in appearance and cell growth with the basic processes of cell division. Determine if cell division is always helpful or sometimes harmful.

1 — Have the students bring childhood pictures of themselves to class. Post them by number.

Ask them to remember as much as they can about how they grew...how their own features changed.

Ask the students to try to identify their classmates from the pictures. List the ways they have changed.

In groups of five, compare "how you did in the identification process. Tell your classmates how you recognized them."

LEFT MODE

RIGHT MODE

3 — Figure out a way to show the difference between mitosis and meiosis, using full body movement activity. The whole class must participate in the "dance."

Do a series of calculations on how many cells would be present after ten cycles of division, 100 cycles, 1000 cycles. Check their knowledge of the differences and similarities between mitosis and meiosis.

2 — "Gather pictures from your own sources of from magazines that show how other animals grow and change." Ask the students to compare their own growth changes with other species.

Lecture on and assign readings on how and why body cells divide.

Pay particular attention to relating cell division to growth and change through time.

LEFT MODE

LEFT MODE

LEFT MODE

RIGHT MODE

This lesson was paraphrased from a science curriculum plan. It is a typical text assignment. We have rounded it out in a 4MAT fashion.

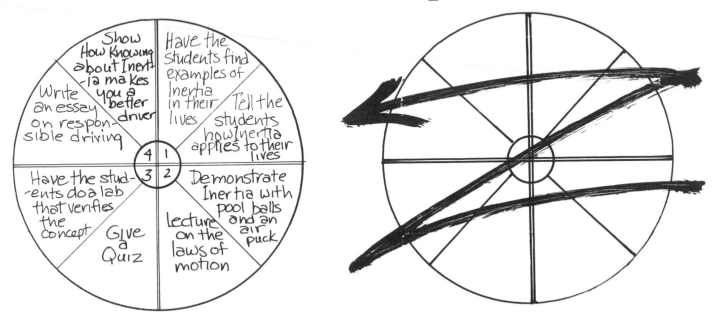

This is a classic science teacher variation on the standard learning cycle. Here the teacher begins in Quadrant Two, then moves to Three, then to One, and finally to Four.

This is not intended as a parody, for we have all seen this instructional cycle in classrooms the country over. The problem is simple . . . the students do not get concrete experience, (true Quadrant One and Four). Instead they get abstract experience **about** concrete experience. Too often teachers provide a cursory statement about their own values, their own concrete experience.

THIS IS NOT 4MAT! WE MUST INVOLVE THE STUDENTS' VALUES AND THE CONCRETE EXPERIENCE OF THE STUDENTS.

As should be clear by now, our bias is to teach **4MAT** units starting in Quadrant One and progressing in sequence through Quadrant Four with explicit attention paid to Right/Left Mode techniques. Experience has shown that the Quadrant One Learners are the easiest to lose in teaching—thus our choice to serve them first. But habits run deep. Many teachers we meet in workshops do not have the reflex to start with personal meaning and concrete experience.

Science teachers have traditionally focused their basic instruction and enrichment activities in Quadrants Two and Three. Over the years we have had to coach science teachers differently than teachers of writing, art, and social studies. With the tendency to start with abstract concepts so strong among science teachers, we suggest the following.

If you feel you must start in Quadrant Two or Three, we strongly suggest that you attempt to go from your chosen starting place, and then proceed completely around the cycle and back to your beginning point. But do not exit the learning cycle here. Instead we urge you to continue through the cycle and exit from Quadrant Four.

The practice of continuing onward through the cycle honors a basic tendency so often overlooked, the tendency to complete the exploration of a concept within the context of life skills. Quadrant Four insures this.

In **4MAT**, enrichment is holism. It is an exploration of a concept through Learning Styles and Learning Modalities. In practice, many teachers attend to enrichment by means of a return trip through the Quadrants once the first cycle is complete. They go again into Quadrant One concerns after Quadrant Four's synthesis and application have been explored. This "re-cycle", and its commitment to the students' personal preferences and to the wide variety of facets of the concept, is an excursion into integrity.

Too often "enrichment" is narrowly confined to Quadrant Two and Quadrant Three approaches. If enrichment serves to honor how what students learn can be applied in their lives, one of the products of the educational process will be a passion for lifelong learning. Perhaps then students will sense that education has been an experience designed to serve their lives.

138

Secrets From Real Scientists

The image of the scientist in the afore mentioned survey of elementary school students (the white-coated male working at a microscope) was a fairly bleak stereotype. Such a view falls far short of depicting the actual work patterns of most scientists. The basic tool of the scientist, the human brain, possesses attributes of discipline and rationality coupled with imagination and creativity. The authors of this book continue to be amazed at how few imagination and creativity activities are written into most science texts.

Because there is a passion for exactness and precision in the scientist's world, many miss the sense of excitement born of the quest for the unknown. Scientists are ever alert to the variety of sources of their awareness. Many know that ideas come to their attention in ways that defy explanation. Some appear in the labyrinth of a dream or come popping into their minds as they put on their shoes. Nearly all scientists have some sort of method to capture these fleeting thoughts.

Beyond the very necessity of capturing ideas whenever they appear, scientists also have unique ways of shifting consciousness so as to sense more than might be expected. They have methods for letting their minds become, in spirit, the creatures they are studying. Dr. Jane Goodall is a fine example of this. Others allow their minds to become the inner workings of a machine or system. Still others turn their minds into energy and follow the flow of a process from start to finish.

The following explores some of the methods scientists use to better understand the worlds they study.

Empathizing

Empathy is the projection of one's own personality into an object, with the attribution to the object of one's own emotions and responses.

The scientist extends this definition into another dimension. Whereas empathy, as defined, displaces the viewer's personality, the scientist wishes more to adopt the persona of the object. Dr. Jonas Salk, the world renowned immunologist, recounts that when he was a child he would play a game of becoming the objects of his interest. If he were interested in a tree, he would become the tree. His senses felt the roots, the wind in its branches, and the rush of nutrients within its body. Later in life he carried this talent with him to the laboratories where he sought an answer to the crippling and killing disease poliomyelitis.

He visualized himself as a polio virus and attempted to sense how it would reproduce, spread and sustain itself. At the same time he became the body's immune system and tried to sense how it would attempt to repel the intruder. Sometimes he became an antigen, other times a white blood cell, and so on. Eventually the patterns became clear and the life saving vaccine for polio was created.

Many times helping students explore the process of empathizing results in unique insights. (This procedure is particularly valuable with Quadrant One students.) Empathizing opens up whole new vistas as one becomes a pulley, a magnet, an osmotic membrane or an endangered or even extinct species.

Quantifying And Estimating

Quantification and measurement skills are basic to every science. Nearly all the curriculum projects of the 1960s and 70s were heavily commited to helping students develop these skills. Of particular note were the techniques developed by project KARE.[19] Students were asked to estimate how many leaves were on a tree, cars in a parking lot, blades of grass in a lawn, rocks in a planter, bricks in a wall. The assigned tasks could be counted and so comparisons could be made between the estimates and the actual count.

Using this technique, teachers found that students' skills could be honed to a remarkable degree of accuracy and precision. It is delightful fun to ask students to produce problems for the teacher or even have estimation derbies. These can be given as homework assignments or as special projects.

Quadrant Three Learners have shown remarkable aptitude with activities involving this form of "guesstimating." Eventually large scale projects can be attempted such as the classic assignment from ESSENCE[20], " Go outside and find a million of something and prove it." The handling of large quantities of phenomena is a demand in most of the sciences. This kind of estimation and quantification is an overriding gestalt in the sciences. Those who are skilled at these tasks are automatically at an advantage in a laboratory and in the field.

Dreaming

Many of the most important insights leading to innovation have come to scientists in their dreams. Einstein dreamed he rode a beam of light and the experience eventually led him to his theory of relativity. The formulation of the benzene ring in organic chemistry came to Kekule in a dream. Thomas Edison was so enchanted by the contributions his dreams made to his work that he kept a cot in his laboratory at all times.

The subject of dreams has been the center of much controversy. There are those who believe dreams are little more than useless information the brain is flushing away. Others, and by far the majority, are able to document the real contribution that dreams make to their work in fields as varied as film-making, medicine, engineering, and management.

Dream harvesting is a skill just like any other skill in scientific perception. Some have formulated natural techniques for remembering their dreams, while others have trained themselves to follow a series of steps in the process. We have had success with dream harvesting by students as young as three years and as old as eighty four.

The process of dream study is simple. Some preparation is needed. People are doubtful initially that they can remember their dreams. It is helpful to initiate a short discussion of dreams since students have ideas about them and are interested in sharing them.

It is helpful to pick a concept the students are studying such as osmosis, evolution, mass, energy, mammals, etc. Ask the students to review all the things they can recall about the concept and write a short paragraph about it. They keep the paragraph and re-read it to themselves prior to going to sleep. Then they follow this procedure:

1.) Place a pad and pencil next to the bed.

2.) Read the concept paragraph and then close your eyes and visualize all you can about the concept.

3.) Repeat to yourself that at the end of dreaming you will wake up.

4.) Go to sleep. . .dream . . .and wake up at the end of your dream.

5.) When you wake at dream's end, KEEP YOUR EYES CLOSED. Review the dream and write down a few phrases about the dream.

6.) Go back to sleep.

7.) In the morning re-read your dream notes and try to relate the content to the concept you were trying to dream about. Be attentive to physical details.

At first the students may be frustrated, but do persist. Although they may not make any direct connection with the chosen concept, they eventually will. As people become more facile in the act of paying attention to dreams, they develop a high capacity to seek specific information and relationships. The more skilled become sensitive to nuance and subtleties that enrich and enhance perception.

Creative Dramatics

Perhaps the highest form of synthesis of the various learning modalities discussed earlier is creative dramatics. Students can gain remarkable understanding of the basic elements of a concept, including its more tacit meanings, through high expression in visual, auditory, and kinesthetic expressions.

One way to approach creative dramatics is to have one or more students pick a concept that is being studied and develop a way of expressing that concept back to the class. The students may not use any words, numbers, or any other standard form of communication. They are permitted any form of movement and any kind of sound that communicates the concept. Sometimes rehearsal time is needed so the beginning of the assignment may be given for homework. Teachers need to help students develop multiple ways of expressing what they learn.

We want the students to be as adept in expressing the felt meaning of an idea as its abstract meaning. If creative dramatics is used properly, the entire cycle of student Learning Style preferences can be employed.

Quadrant One students add the personal meaning,
Quadrant Two students provide the accuracy,
Quadrant Three students keep the action consistent with the idea,
Quadrant Four students choreograph the effort for presentation to the rest of the class.

The Arts And Science

Research into preferred ways of receiving information shows individuals have different modality preferences. Some prefer visual experiences, some kinesthetic, and some seek auditory input. Schools demand that expressions of learning be made in writing, mathematics, or in standard speech. In schools where modality alternatives have been used, students show amazing improvement in both scholastic performance and self esteem.[21]

One of the best ways to invite students to explore scientific concepts through the arts is to ask them to do homework in which they blend the arts and the sciences.

For example, "Draw or paint a portrait of gravity," or

"Sculpt a redesign of the human face," or

"Write a song that tells us what the water cycle is like."

"Create a dance that illustrates osmosis."

"Make a poster that celebrates Newton's Laws," or

"Paint a portrait of the birth of an island."

Also have art materials become an integral part of classroom supplies so that they can be drawn upon as readily as the science materials. For example, have students take notes only on the right side of their open notebooks, saving the left sides for drawings or images that illustrate the concepts being recorded.

Spontaneous Monologue

This is a technique developed by Richard Herrmann[22] at the University of Vermont. It is a close partner to dream harvesting. It requires that the students go through a series of steps that attempt to extend their domain of knowing systematically.

1. Ask the students to write down three things that are important to know about the concept being studied.

2. Then suggest they write down three facts about each of the three things written in Step One.

3. Ask the students to pick any one of the first "important" things with its list of three facts, and write down their feelings related to the concept.

4. Invite them to get the appropriate materials and draw or paint an image that expresses the concept.

5. Then ask the students to get paper and a pencil handy on their desks.

6. Have them sit back, relax, and close their eyes. Then ask them to visualize everything their minds contain about the concept. Encourage them to be sensitive to color, odors, textures, relationships, motion, etc. Allow about two minutes for this.

7. Have the students then open their eyes and write in a "free flowing" form, with no attention to spelling, grammar and syntax, whatever comes to mind. "Whatever flows, goes."
Allow about five minutes for this.

8. Then have the students re-read what they have written and make any changes they feel is necessary. Once this is done, the teacher may choose to ask them to share their efforts in small groups or to the entire class. It may be appropriate just to collect the papers and read the variety of insights they contain. We do not recommend grading these papers.

A variation on this strategy is to substitute real world experience for steps One through Four. This means a laboratory session, a field trip, or even some commonly shared experience like a community event can replace the intellectual excursion described above.

Example: Have two students, one being blindfolded, work together setting up a microscope. Obviously one student must guide the other through the process. The task is to have the blindfolded student set up a wet slide preparation. The sighted student must describe the images that appear in the microscope. Both students will have very real experiences, but they will be totally different. Once completed, repeat steps Five through Eight above.

This technique is often used by Jonas Salk in his journal work. Two of the authors of this book, Bill Hammond and Bob Samples, use this method extensively in their own journal keeping.

In all the work of reputable scientists and writers that we have been privileged to examine, the activities similar to spontaneous monologue invariably show the highest risk and originality of thought. It is imagination expanding, it is pattern seeking, it is amazing at its best.

It is a steppingstone to competence.

One of the last things written by Gregory Bateson, the noted scientist who made significant contributions in many fields, was that metaphor supplied the foundation for all science. He mused in a Lindisfarne Lecture[23] that science was all metaphor, and it was a metaphoric process involving verbs rather than nouns.

Bob Samples, one of the authors of this book, explored the metaphoric functions of the brain-mind system in **The Metaphoric Mind.**[24] The processes of metaphor provide a linkage between seemingly unrelated things. Many of the activities in various instructional approaches in Learning Modalities and Learning Styles result in students creating metaphoric bridges between themselves and the concepts they study. Don and Judy Sanders in their book, **Teaching Creativity Through Metaphor: An Integrated Brain Approach** comment on the importance of the metaphor:

"Indeed, metaphors are tools for insight, for understanding. It is with the metaphor that we can unlock the part of our minds that schools have traditionally left closed and untapped, the part of our minds where conceptual imagery resides, the right hemisphere."[25]

As students gain in metaphoric maturity they learn the processes involved in making connections. They also learn a great deal about the interconnectedness of ideas. When we vary learning approaches through the use of Learning Modalities and Learning Styles we feed the origins of metaphoric thought. We facilitate the construction of relevant bridges between learners and concepts. Consider the following activity:

1. Give a team of six or more students the use of a chalkboard or a ten foot length of "butcher paper". The paper is recommended as it provides a lasting record.

2. Begin with a single concept. Write it on the paper or chalkboard.

3. Invite the students to begin calling out what comes to mind when they consider the concept.

4. Write their offerings in columns. Take care not to edit or comment.

Metaphoring the concept of Gravity

pull
heavy
tug
lift
force
mass
weight
down
moon
oceantides
fall
push
mass
Newton
apple
spaceship
rockets
landslide
orbit
jump
grasp
acceleration
earthworm
buick
mother hubbard

humpty
dumpty
lace
elbows
serious
heavy
tidal
bridal
surf
surfsup
tree
swing
fling
bring
Montana
lineman
throw
pass
riot
healthcare
shadow
muddylace
downside
insidejob
froth

dogfood
fat
flower power
mountain top
candy apple
applesauce
dirty looks
Mr. Jackson (the principal)
sand in your sneakers
corn-on-the-cob drag
moon dust
Mr. Spock
dandelions
granola
bumblebee
rocketree
down flower
apple power
orbitsauce
big newton
spock talk
logicat
downtide
tugtide
beagle puppy
beaglesneakers

downpower
weightworm
humblebee
moonlace
corndown
downflakes
sinkpower
fill down

sinking
feeling
downfeel
leadballoon
thrust us

In gravity we
trust

(Here is a poem that resulted from step six.)

Power to the push-pull

Who is the push?
Who is the pull?
Is it from above,
Or does it live below?

Newton had his apple
and saw it pulled right down.

But Humpty had his dumpty
and saw it turned around!

Push that apple
Pull that apple

How can we chase it down
When Sir. Isaac won't talk
to eggs?

This example is from "Are You Teaching Only One Side of the Brain?" by Bob Samples in Learning Magazine, February, 1975, page 27.

5. Continue until forty to eighty offerings are recorded.

6. Ask the students to write a statement about the original concept, using the list of words that has been created as the primary resource.

7. Share the statements and encourage the students to explore the diversity that emerges.

Another variation is to have all the students draw or paint an image suggested by the concept being studied. The images can then be displayed in a "visual metaphor gallery." In the discussion that follows as the students comment on the "messages" in the art the linkages and common themes can be explored. This variation is useful if the students either cannot or are hesitant to write.

Verbal techniques can also encourage metaphoric thinking. The Sanders include many of these in their book.   Below are examples. [25]

If dinosaurs are grandfather clocks and ancient cities are sundials, what clock symbolizes modern American society?

Science is a process that "tells time' by exploring nature. Show this to be true.

How do trees tell time? Rivers tell time? Rocks tell time? Birds tell time?

The ESSENCE materials also contained many examples. [20]

Go outside and find indirect evidence of a population of something.

Go outside in the community and find who are predators and who are prey.

Go outside and relate the following pairs of words.

> garbage can/stomach
> kid/number
> lifetime/word
> school/supermarket
> time/being hungry
> love/a water fountain

Concept Mapping

Concept mapping is a technique used to explore underlying relationships. It is partly visual in a graphic sense, and partly metaphoric. The methods were first explored by the Nuffield Foundation science materials in the late 1950s. They eventually found their way into the science programs of the Ontario Ministry of Education in the 1960s. David Hawkins was the American champion of this approach in 1961. He was the was director of the Elementary Science Study group in Cambridge, Massachusetts at that time.

Literally thousands of class hours were spent by students at all levels gaining insights into the inter-relatedness of objects, events, and processes occurring in the natural and social world. This is not only a technique of science but of learning in general. Tony Buzan[26], a creativity specialist in England, believes the approach is central to creativity. Gabrielle Rico in her book, **Writing the Natural Way,** cites its usefulness in teaching and developing grace in writing.[27]

The process consists of free flowing thoughts that spark ideas as one moves from a central point. It is essentially the same process described in the metaphor activity earlier. It is a process of tapping past experiences and evoking feelings and thoughts in a stream of consciousness fashion with amazing results.

As the process unfolds, the connections are made between and among seemingly unrelated ideas. These "irrational connections" are a source of wonder to the writer. An example of a concept map is shown below.

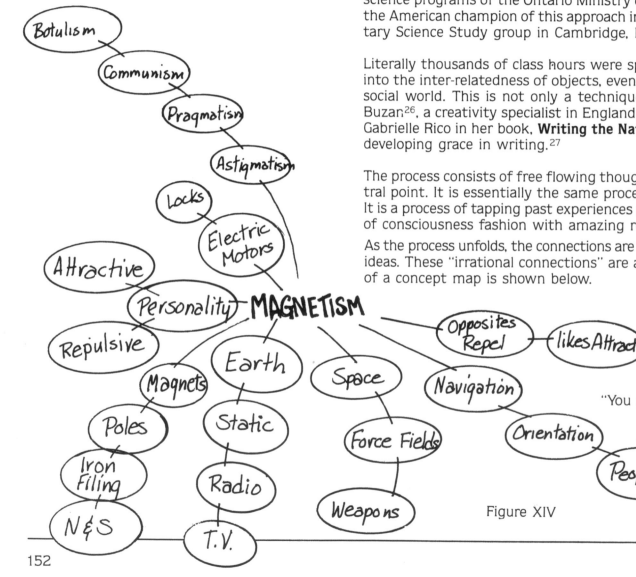

"You mean I knew/thought/hunched/felt all that!"

Figure XIV

A more formal kind of concept mapping is promoted by Joseph Novak and D.Bob Gowin in their new book, **Learning How to Learn**.[28] Their approach focuses on how student-originated concept maps can be organized to communicate the concept hierarchies accepted in science. This is how one of their "maps" turned out.

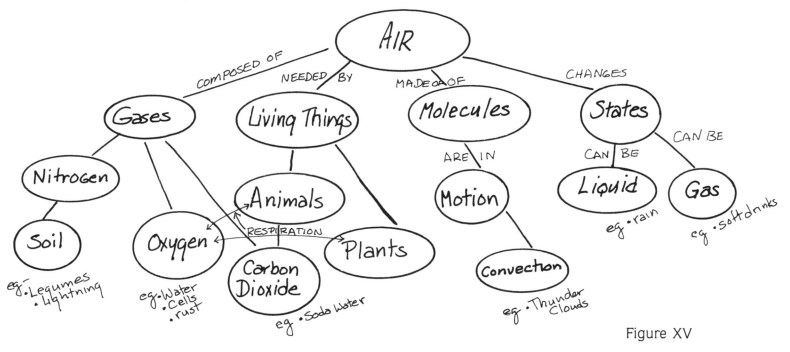

Figure XV

Displacement Characteristics

Nearly all children and a remarkable number of adults are involved some time or other in a Walter Mitty world of displacement. This is what happens when you become some other person. It is a surprise to many to discover that scientists also are Walter Mittys. They report they often use the technique to debate with someone else, or to write a criticism of someone else's work. And as reported earlier in the Jonas Salk example, they often become some object, event, or process in nature.

A successful technique is to provide a foil for displacement. In **The Tao of Pooh**[29] by Peter Hoff, it is clear that Pooh is a perfect displacement figure. Children take remarkable risks when they feel the safety of speaking through another. The displacement foil can offer insights, suggestions, or recommendations, and all from a neutral position. Try the following:

What would Pooh say about how a rock thinks?

How would Pooh explain GRAVITY?

Have Pooh explain what a predator thinks when it sees its prey.

The displacement can be done in two ways. One is the displacement of human attributes to other objects, events, and processes. The second is displacement of the natural attributes of objects, events, and processes to humans. The process is important as it allows reticent students to take risks.

We began this section with the idea that practicing scientists use many interesting techniques to spark creativity. We have listed a few. It is our hope that teachers and students will find them entertaining as well as useful. These techniques are an integral part of the authors' lives and we attest to how richly they have contributed to our work.

Science is an anthem to the human mind. It has been shrouded in an elitist mystique for too long. We can no longer afford such a vision.

Science is not an activity reserved for the coldly detached, methodically objective.

Science has an integrity born of human beings. It is an artform in which accuracy, precision, and responsibility are prized, but it is not heartless and uncaring.

Conclusion

There was a period of time when science pretended to be immune from the consequences of its discoveries.

But those days melted into the vapor clouds of Hiroshima and Nagasaki. The parched spread of the Sahara and other world deserts are as much the result of science as are the hoped-for reversals and solutions to ecological devastation.

Science has grown to a mature, responsible, compassionate realm of human commitment. With this maturity it is more likely that young people will be drawn again into seeking partnership between themselves and the cosmos. The instructional approaches expressed in this book may hopefully facilitate such bonding. There is a growing possibility that science will be seen as the place where the likes of R. Buckminster Fuller, Thomas Edison, Margaret Mead, Einstein and Gregory Bateson inspired the most exciting generation of scientists the world has ever known. Some cited in this book are turning our biases and stereotypes about science into a rich and humane vision that is filled with hope and honesty. There are poignant voices being heard among today's scientists to honor the diversity that science represents.

Our hope is that these pages encourage a recognition of that diversity and its integrity. We want young people to love science—for science is the way we have invented to unite our minds with all that is. It is a union that is born of the grand sweep of human capabilities standing in awe of the world.

Science is our own creation. If there is to be joy in science, the joy must be born within us. If there is to be integrity in science, there must be integrity within us. Both joy and integrity are woven into the fiber of human wholeness. It is this wholeness we are committed to serve.

Who The Authors Are

Bob Samples is an independent scholar and author. He is the international director of the Solstice Seminars. In the past he served as a writer/consultant and staff person with the Elementary Science Study, Man: A Course of Study, The Earth Science Curriculum Project, and Environmental Studies for Urban Youth, all funded by the National Science Foundation. Currently he is involved in studying the interrelationship between the brain-mind and natural systems with particular attention to the origins of creative thought. The results of this work are being applied to education and management. Bob is popular as a keynote speaker, lecturer, and workshop leader throughout the world.

Bill Hammond is Director of Environmental Education and Instructional Development Services for the Lee County Schools of Fort Myers, Florida. He has been a science teacher, supervisor, and curriculum innovator for more than twenty years. He has served as a science consultant to the American Association for the Advancement of Science, the National Science Teachers Association, Project Learning Tree, and Project WILD. Bill is a student of natural systems and has developed implementation models for long term change. There are great demands for his speaking, workshop, and seminar appearances. He is a trainer for Applied Creative Thinking and serves as an officer and board member of several nationally known groups with environmental concerns.

Bernice McCarthy is founder and Director of EXCEL incorporated. In 1979 she was awarded a grant by the McDonald Corporation to host an innovative conference comprised of leading researchers in the neurosciences, psychology, and education. As a result of this conference, Bernice created **The 4MAT System**. Currently she is exploring the application of this holistic approach in management and other fields. She has twenty seven years of teaching experience at all levels from kindergarten through university. She is widely sought as an inspirational speaker and workshop leader.

1. Dewey, John. *Democracy in Education*. New York: Holt, 1913.

2. Course Curriculum Improvement Projects funded by the National Science Foundation. Published by a variety of publishers.

3. Rowe, Mary Budd. "Science Education: A Framework for Decision-Makers." *Daedelus*. Vol. 112, No. 2, pp. 123-142.

4. Sperry, Roger W. "Lateral Specialization of Cerebral Function in the Surgically Separated Hemispheres." *The Psychophysiology of Thinking*. F.J. McGuigan and R.A. Schoonover, Eds. New York: Academic Press, 1973.

5. Bogen, Joseph E. "The Other Side of the Brain: An Appositional Mind." *Bulletin of the Los Angeles Neurological Societies*. Vol. 34, July 1969.

6. Samples, Bob. "Mind Cycles and Learning." *The Kappan*. Vol. 58, No. 9, May 1977, p. 688.

7. Kolb, David. *Experiential Learning: Experience as a Source of Learning and Development*. Englewood Cliffs, New Jersey: Prentice Hall, 1983.

8. McCarthy, Bernice. *The 4MAT System: Teaching to Learning Styles with Right/Left Mode Techniques*. Barrington, Illinois: EXCEL, Inc., 1980.

9. *Ibid*.

10. Streufert, Siegfried. *Style of Thinking Not I.Q., Tied to Success*. Reported by Daniel Goleman. *New York Times*. July 31, 1984.

11. Schachter, S.; and J.E. Singer. "Cognitive, Social, and Physiological Determinants of Emotional State." *Psychological Review*. 69:379-399. 1962.

12. Novak, Joseph D., and D. Bob Gowin. *Learning How to Learn*. New York: Cambridge University Press, 1984.

13. Rowe, Mary Budd; *op. cit.*

14. Bruner, Jerome S. *The Process of Education*. Cambridge, Massachusetts: Harvard University Press, 1960.

15. Holdzkom, David; and Pamela Lutz. *Research Within Reach: Science Education*. Washington, D.C.: National Institute of Education, 1984.

16. Otto, James H., and Albert Towle. *Modern Biology*. New York: Holt, Rinehart, and Winston, 1973.

17. Brandwein, Paul F.; Warren E. Yasso; and Daniel B. Rovey. *Life: A Biological Science, Concepts in Science*. New York: Harcourt, Brace, and Jovanovich, 1980.

18. Mead, Margaret. Conversation with one of the authors.

19. *KARE*. Erdenheim, Pennsylvania: Montgomery County Intermediate District Unit 23.

20. *Essence: Environmental Studies for Urban Youth Project Materials*. Reading, Massachusetts: Addison Wesley, 1972.

21. *Project Viewpoint Summary Report*. Gresham, Oregon: Union High School District, E.S.E.A. Title III Project. Susan Miller, Principal Investigator.

22. Herrmann, Richard. Workshop presented at the Second Whole Brain Symposium. Key West, Florida, 1982.

23. Bateson, Gregory. *Tape to the Lindisfarne Fellows*. West Stockbridge, Massachusetts: Lindisfarne Press, 1980.

24. Samples, Bob. The Metaphoric Mind. Reading, Massachusetts: Addison Wesley, 1976.

25. Sanders, Donald A.; and Judith A. Sanders. *Teaching Creativity Through Metaphor*. New York: Longman, 1984.

26. Buzan, Tony. *The Evolving Brain*. New York: Holt, Rinehart and Winston.

27. Rico, Gabriele Lusser. *Writing the Natural Way*. Los Angeles: J.P. Tarcher, 1983.

28. Novak, Joseph; and D. Bob Gowin, *op. cit.*

29. Hoff, Peter. *The Tao of Pooh*. New York: Penguin Press, 1982.

Bibliography

Bateson, Gregory. *Mind and Nature.* New York: E.P. Dutton, 1979.

Berman, Morris, *The Reenchantment of the World.* Ithaca: Cornell University Press, 1981.

Bloom, Floyd E.; Arlyne Lazerson; and Laura Hofstader. *Brain, Mind and Behavior.* New York: W.H. Freeman and Company, 1985.

Bruner, Jerome S. *On Knowing.* New York: Atheneum, 1966.

Charles, Cheryl; and Bob Samples, Eds. *Science and Society.* Washington, D.C.: National Council for the Social Studies, 1978.

Dohemann, Warren; and Melvin Suhd. *The Alchemy of Intelligence.* Lake Oswego, Oregon: Metamorphous Press, 1984.

Donaldson, Margaret. *Children's Minds.* Glasgow: Wm. Collins Sons, 1978.

Dychtwald, Ken. *Bodymind.* New York: Pantheon Books, 1977.

Fuller, R. Buckminster. *Critical Path.* New York: St. Martins Press, 1981.

Gardner, Howard. *Frames of Mind.* New York: Basic Books, 1983.

Hampden-Turner, Charles. *Maps of the Mind.* New York: Macmillan, 1981.

Hawkins, David. *The Informed Vision.* New York: Agathon Press, 1974.

Jones, Richard. *Fantasy and Feeling in Education.* New York: New York University Press, 1968.

Knapp, Clifford E.; and Joel Goodman. *Humanizing Environmental Education.* Martinsville, Indiana: The American Camping Association, 1983.

McKim, Robert. *Experiences in Visual Thinking.* Monterey, California: Brooks/Cole, 1972.

Mead, Margaret. "Creating a Scientific Climate for Children." *Science and Children.* May, 1977.

Prigogine, Ilya; and Isabelle Stengers. *Order out of Chaos.* New York: Bantam Books, 1984.

Restak, Richard M. *The Brain.* New York: Bantam Books, 1984.

Salk, Jonas. *Anatomy of Reality.* New York: Columbia University Press, 1983.

Samples, Bob. "Holonomic Knowing." *Education in the 80's.* R. Baird Schuman, Ed. Washington, D.C.: National Education Association, 1981.

Samson, Richard W. *Thinking Skills.* Stamford, Connecticut: Innovative Sciences, Inc., 1981.

Thomas, Lewis. *Late Night Thoughts on Listening to Mahler's Ninth Symphony.* New York: Viking, 1983.

Wilson, Edward O. *Biophilia.* Cambridge, Massachusetts: Harvard University Press, 1984.

Whitrock, M.C., Ed. *The Brain and Psychology.* New York: Academic Press, 1980.